Wilson Flagg

**The birds and Seasons of New England**

Wilson Flagg

**The birds and Seasons of New England**

ISBN/EAN: 9783744719612

Printed in Europe, USA, Canada, Australia, Japan

Cover: Foto ©berggeist007 / pixelio.de

More available books at **www.hansebooks.com**

THE

# BIRDS AND SEASONS

OF

# NEW ENGLAND.

By WILSON FLAGG,

AUTHOR OF "THE WOODS AND BY-WAYS OF NEW ENGLAND."

With Illustrations.

"I walk the hills my feet first knew, and from year to year they grow dearer and dearer to me." — R. A. OAKES.

BOSTON:
JAMES R. OSGOOD AND COMPANY,
LATE TICKNOR & FIELDS, AND FIELDS, OSGOOD, & Co.
1875.

TO

## SAMUEL P. FOWLER, Esq.,

THE AMATEUR NATURALIST;

THE FRIEND AND ABLE ADVOCATE OF THE BIRDS,

AND

THE USEFUL AND ENTERTAINING WRITER,

This Volume

IS RESPECTFULLY INSCRIBED BY ONE OF THE ADMIRERS
OF HIS GENIUS AND HIS WORTH,

THE AUTHOR.

# INTRODUCTION.

THE title of this work does not give the reader a full understanding of its scope and contents, as it treats of Scenes and Flowers as well as of Birds and Seasons. Its present form was adopted for the sake of brevity. My classification of Birds is wholly arbitrary, but not without signification. In the Index I have given their scientific names, chiefly according to Nuttall, preferring those which were used by our early writers on Ornithology, because the species can be more easily identified by those than by the Greek names applied to them in the new nomenclature.

My essays are not biographies of the Birds. I treat of them chiefly as songsters, and speak only of those habits which render them useful, interesting, or picturesque. I have confined myself principally to my own personal observations, but have freely quoted from several authors. I ought to remark in this place that I am much indebted to Mr. John Burroughs, whose essays on Birds and kindred subjects in "The Atlantic Monthly" I formerly read with great pleasure.

Mr. Burroughs is confessedly the most graphic and entertaining of our authors on Ornithology. I regret that I had not seen his book, "Wake Robin," before this volume was in type, as the perusal of it would have improved my own pages.

I would remind the reader that some parts of my book have previously appeared in print.

WILSON FLAGG.

# ILLUSTRATIONS.

[Printed by the Heliotype Process, after Views from Nature.]

|  | PAGE |
|---|---|
| THE OLD HOMESTEAD OF GENERAL PUTNAM, IN DANVERS | *Frontispiece.* |
| VIEW OF CHARLES RIVER FROM THE FARM OF MR. ANTHONY HOLBROOK, IN AUBURNDALE | 43 |
| SCENE NEAR FRESH POND, IN CAMBRIDGE, ON CONCORD TURNPIKE | 83 |
| VIEW OF NEPONSET RIVER IN MATTAPAN | 118 |
| VIEW IN LYNN, LOOKING THROUGH A VISTA OF TREES ON A DESCENDING ROAD | 164 |
| OLD ROAD, AS SEEN FROM THE HILL LEADING TO SPOT POND, ON THE ROUTE FROM MEDFORD TO STONEHAM | 202 |
| SCENE ON BASS RIVER, IN RYALL SIDE, BEVERLY | 248 |
| MILL SCENE IN BOXFORD | 277 |
| VIEW OF LANDSCAPE, INCLUDING THE SOUTH SIDE OF CAPTAIN ENOCH WOOD'S ESTATE, IN WEST BOXFORD | 333 |
| VIEW OF OLD WINDING ROAD IN NORTH ANDOVER | 364 |
| "DEN ROCK," IN ANDOVER, ON THE OLD SALEM TURNPIKE | 403 |
| VIEW OF BIRCH BROOK, IN LYNN | 431 |

# THE BIRDS AND SEASONS OF NEW ENGLAND.

## MUSIC OF BIRDS.

AMONG civilized people those are the most cheerful and happy, if possessed of a benevolent heart and favored with the ordinary gifts of fortune, who have acquired by habit and education the power of deriving pleasure from the objects that lie immediately around them. But these sources of happiness are open to those only who are endowed with sensibility, and who have received a favorable intellectual training. The more ordinary the mental and moral organization and culture of the individual, the more far-fetched and dear-bought must be his enjoyments. Nature has given us in full development only those appetites which are necessary to our physical well-being. She has left our moral powers and affections in the germ, to be developed by education and reflection. Hence that serene delight that comes chiefly from the exercise of the imagination and the moral sentiments can be felt only by persons of superior and peculiar refinement of mind. The ignorant and rude are dazzled and delighted by the display of gorgeous splendor, and charmed by loud and stirring sounds. But the more simple melodies and less attractive colors and forms, that appeal to the imagination for their principal effect, are felt only by individuals of a poetic temperament.

In proportion as we have been trained to be agreeably affected by the outward forms of nature and the sounds that proceed from the animate and the inanimate world are we capable of being happy without resorting to vulgar and costly recreations. Then will the aspects of nature, continually changing with the progress of the seasons, and the songs that enliven their march, satisfy that craving for agreeable sensations which would otherwise lead us away from humble and healthful pursuits to those of an artificial and exciting life. The value of these pleasures of sentiment is derived not so much from their cheapness as from their favorable moral influences, that improve and pleasantly exercise the mind without tasking its powers. Those quiet emotions, half musical and half poetical, which are awakened by the songs of birds, belong to this class of refined enjoyments.

But the music of birds, though delightful to all, conveys active and durable pleasure only to those who have learned to associate with their notes, in connection with the scenes of nature, a crowd of interesting and romantic images. To many persons of this character it affords more delight than the most brilliant music of the concert or the opera. In vain will it be said as an objection, that the notes of birds have no charm save that of association, and do not equal the melody of a simple reed or flageolet. It is sufficient to reply that the most delightful influences of nature proceed from sights and sounds that appeal to a poetic sentiment through the medium of slight and almost insensible impressions made upon the eye and the ear. At the moment when these physical impressions exceed a certain mean, the spell is broken, and the enjoyment, if it continues, becomes sensual, not intellectual. How soon, indeed, would the songs of birds pall upon the ear if they were loud and brilliant like a band of instruments. It is simplicity that gives them their charm.

As an illustration of the truth of this remark, I would say that simple melodies have among all people exercised a greater power over the imagination, though producing less pleasure to the ear, than louder and more complicated music. Nature employs a very small amount of physical agency to create sentiment, and when an excess is used a diminished effect is produced. I am persuaded that the effect of our sacred music is injured by an excess of harmony or too great a volume of sound. A loud crash of thunder deafens and terrifies, but its low and distant rumbling produces a pleasant emotion of sublimity.

The songs of birds are as intimately allied with poetry as with music. "Feathered Lyric" is a name that has been applied to the Lark by one of the English poets; and the analogy is apparent when we consider how much the song of this bird resembles a lyrical ballad in its influence on the mind. Though the song of a bird is without words, how plainly does it suggest a long train of agreeable images of love, beauty, friendship, and home! When a young person is affected with grief, he seldom fails, if endowed with a sensitive mind, to listen to the birds as sympathizers in his affliction. Through them the deities of the grove seem to offer him their consolation. By his companionship with the objects of nature all pleasing sights and sounds have become anodynes for his sorrow; and those who have this mental alembic for turning grief into poetic melancholy cannot be reduced to despondency. This poetic sentiment exalts our pleasures and soothes our afflictions by some illusive charm, derived from religion or romance. Without this reflection of light from poetry, what is the passion of love, and what our love of beauty, but a mere gravitation?

The music of birds is modulated in pleasant unison with all the chords of affection and imagination, filling the soul with a lively consciousness of beauty and delight. It soothes us with romantic visions of love when an ethereal sentiment of adoration as well as a passion, and of friendship when a passion and not an expediency. It reminds us of dear and simple adventures, and of the comrades who had part in them; of dappled mornings and of serene and glowing sunsets; of sequestered nooks and of sunny seats in the wildwood; of paths by the waterside, and of flowers that smiled a bright welcome to our rambling; of lingering departures from home, and of old by-ways hedged with viburnums and overshadowed by trees that spread their perfume around our path to gladden our return. By this pleasant instrumentality has Nature provided for the happiness of all who have learned to be delighted with her works, and with the sound of those voices which she has appointed to communicate to the human soul the joys of the inferior creation.

# BIRDS OF THE GARDEN AND ORCHARD.

## I.

THE singing-birds whose notes are familiar to us in towns and villages and in the suburbs of cities are strangers to the deep woods and solitary pastures. Our familiar birds follow in the wake of the pioneer of the wilderness, and increase in numbers with the clearing and settlement of the country, not from any feeling of dependence on the protection of man, but from the greater supply of insect food caused by the tilling of the ground. It is well known that the labors of the farmer cause an excessive multiplication of all those insects whose larvæ are cherished in the soil, and of all that infest the garden and orchard. The farm is capable of supporting insects in the ratio of its capacity for producing fruit. These will multiply with their means of subsistence contained in and upon the earth; and birds, if not destroyed by man, will increase with the insects that constitute their food.

Hence we may explain the fact, which often excites surprise, that more singing-birds are seen in the suburbs of a great city than in the deep forest, where, even in the vocal season, the silence is sometimes melancholy. The species which are thus familiar in their habits, though but a small part of the whole number, include nearly all the singing-birds that are known to the generality of our people. These are the birds of the garden and orchard. There are many other species, wild and solitary in their habits, which are delightful songsters in the uncultivated regions lying outside of the farm. Even these are rare in the depths of the forest. They live on the edge of the wood and the half-wooded pasture. The birds of the garden and orchard have been frequently described, and are very generally known, though but little has been said of their powers and peculiarities of song. In the sketches that follow I have given particular attention to the vocal powers of the different birds, and have attempted to designate the part that each one performs in the grand hymn of Nature.

## THE SONG-SPARROW.

The Song-Sparrow, one of our most familiar birds, claims our first attention as the earliest visitant and latest resident of all the tuneful band, and one that is universally known and admired. He is plain in his vesture, undistinguished from the female by any superiority of plumage. He comes forth in the spring and takes his departure in the autumn in the same suit of russet and gray by which he is always identified. In March, before the violet has ventured to peep out from the southern slope of the pasture or the sunny brow of the hill, while the northern skies are liable at any hour to pour down a storm of sleet and snow, the Song-Sparrow, beguiled by southern winds, has already appeared, and on still mornings may be heard warbling his few merry notes, as if to make the earliest announcement of his arrival. He is therefore the true harbinger of spring; and, if not the sweetest songster, he has the merit of bearing to man the earliest tidings of the opening year, and of proclaiming the first vernal promises of the season. As the notes of those birds that sing only in the night come with a double charm to our ears, because they are harmonized by silence and hallowed by the hour that is sacred to repose, in like manner does the Song-Sparrow delight us in tenfold measure, because he sings the sweet prelude to the universal hymn.

His haunts are fields half cultivated and bordered with wild shrubbery. He is somewhat more timid than the Hair-Bird, that comes close up to our doorsteps to find the crumbs that are swept from our tables. Though his voice is constantly heard in the garden and orchard, he selects a retired spot for his nest, preferring not to trust his progeny to the doubtful mercy of the lords of creation. In some secure retreat, under a tussock of moss or a tuft of low shrubbery, the female sits upon her nest of soft dry grass, containing four or five eggs of a greenish-white surface covered with brownish specks. Beginning in April, she rears two and often three broods during the season, and her mate prolongs his notes until the last brood has flown from the nest.

The notes of the Song-Sparrow would not entitle him to rank with our principal singing-birds, were it not for the remarkable variations in his song, in which I think he is equalled by no other bird. Of these variations there are six or seven that may be distinctly recognized, differing enough to be considered separate tunes, but they are all based upon the same theme. The bird does not warble these in regular succession. It is in the habit of repeating one of them several times, then leaves it and repeats another in a similar manner. Mr. Charles S. Paine, of East Randolph, Massachusetts, was, I believe, the first to observe this habit of the Song-Sparrow. He took note, on one occasion, of the number of times a particular bird sang each of the tunes. As he had numbered them, the bird sang No. 1, 21 times; No. 2, 36 times; No. 3, 23 times; No. 4, 19 times; No. 5, 21 times; No. 6, 32 times; No. 7, 18 times. He made the same experiment with a dozen different individuals; and was confident from these trials that each male has his seven songs, or variations of the theme, and they are all equally irregular in the order of singing them.

After reading Mr. Paine's letter, I listened carefully to the Song-Sparrow, in the summer of 1857, that I might learn to distinguish the different tunes, as reported by him. I had never thought of it before; but in less than a week I could distinctly recognize the whole seven, and was convinced that his observations were perfectly correct. It is remarkable that when one powerful singer takes up a particular tune, other birds in the vicinity will follow with the same. These are mostly in triple time, some in common time, while in others the time could not be distinguished. Each tune, however, consists of four bars or strains, sometimes five, though late in the season the song is frequently broken off at the end of the third strain. This habit of varying his notes through so many changes, and the singularly fine intonations of many of them, entitle the Song-Sparrow to a very high rank as a singing-bird.

There is a plain difference in the expression of these several variations. The one which I have marked No. 3 is very plaintive, and is in common time. No. 2 is the one which I have most frequently heard. No. 5 is querulous and unmusical. There is a remarkable precision in the Song-Sparrow's notes, and the finest singers are those which, in the language of musicians, display the least execution. Some blend their notes together so rapidly and promiscuously, and use so many operatic flourishes, that if all were like them it would be impossible to distinguish the seven different variations in the song of this bird.

Whether these tunes of the Song-Sparrow express to his mate or to others of his species different sentiments, and convey different messages, or whether they are the offspring of mere caprice, I cannot determine. Nor have I learned whether a certain hour of the day or a certain state of the weather predisposes the bird to sing a particular tune. This point may perhaps be determined by

BIRDS OF THE GARDEN AND ORCHARD. 9

some future observer, who may discover that the birds of this species have their matins and their vespers, their songs of rejoicing and their notes of complaint, of courtship when in presence of their mate, and of encouragement and solace when she is sitting upon her nest. Since Nature has a benevolent object in every instinct bestowed upon her creatures, it is not probable that this habit of the Song-Sparrow is one that serves no important end in his life and habits. All the variations of his song are given below; and though individuals differ in their singing, the notes will afford a good general idea of the several tunes.

**No. 7. Brilliant.**

NOTE. — The notes marked *guttural* seem to me to be performed by a rapid trilling of these notes with their octave. No bird sings constantly in so regular time as is represented above, and the intervals between the high notes are very irregular. Both the time and the tune are in great measure *ad libitum*.

## THE VESPER-SPARROW.

Soon after the arrival of the Song-Sparrow, before the flowers are yet conspicuous in the meadows, we are greeted by the more fervent and lengthened notes of the Vesper-Bird, poured forth with a peculiarly pensive modulation. This species resembles the Song-Sparrow, but may be distinguished when on the wing by two white lateral feathers in the tail. The chirp, or complaining note, of the Song-Sparrow is louder and pitched on a lower key. The Vesper-Bird is the less familiar of the two, and, when both are singing at the same time, will be seen to occupy a position more remote from the house. In several places they are distinguished by the names of Ground-Sparrow and Bush-Sparrow, from their supposed different habits of placing their nests. I believe, however, that while the Song-Sparrow always builds upon the ground, the Vesper-Bird builds indifferently upon the ground or in a bush.

The Vesper-Bird, of the two species, attracts more general attention to his notes, because he sings a longer though more monotonous song, and warbles with more fervency. His notes resemble those of the Canary, but they are more subdued and plaintive, and have a reedy sound which is not perceptible in the Canary's tones.

This bird is somewhat periodical in his singing habits, confining his lays in some measure to certain hours of the day and conditions of the weather. The Song-Sparrow sings about equally during every hour from morning till night, and the different performers do not always join in concert. This habit renders the little songster more companionable, but at the same time causes his notes to be less regarded than those of the Vesper-Bird, who sings in concert with others of his kind, and at more regular periods.

The Vesper-Bird joins at day-spring with all his kindred in the general anthem of morn, after which he sings occasionally during the day, especially at an hour when it is still and cloudy, but most fervently during the sun's decline until dusk. Hence is derived the name it bears, from its evening hymn, or vespers. There are particular states of the weather that call out the songsters of this species and make them tuneful, as when rain is suddenly followed by sunshine, or when a clear sky is suddenly darkened by clouds, presenting an occasional morn and an occasional even. In this respect these birds are not peculiar, but by singing together in numbers their habit is more noticeable. We seldom hear one of them singing alone. When one begins, all others in the vicinity immediately join him.

The usual resorts of the Vesper-Bird are the hayfields and pastures, from which he has derived the name of Grass-Finch. His voice is heard frequently by rustic roadsides, where he picks up a considerable part of his subsistence; and it is remarkable that this songster more frequently sings from a fence, a post, or a rail than from a tree or a bush. This is the little bird that so generally serenades us during an evening walk at a short distance from the town, and not so near the woods as the haunts of the Thrushes. When we go out into the country on pleasant days in June or July, at nightfall we hear multitudes of them singing sweetly from many different points in the fields and farms.

## THE HAIR-BIRD.

A gentle and harmless little bird, attracting attention chiefly by his tameness and familiarity, chirping at all hours, but without a very melodious song, is the Hair-Bird, belonging to the family of Sparrows, but differing from all the others in many of his habits. He is one of the smallest of the tribe, of an ashen-brown color above and grayish-white beneath. He wears a little cap or turban of velvety-brown upon his head, and by this mark he is readily distinguished from his kindred. Relying on his diminutive size for security, he comes quite up to our doorstep, mindless of the people who are assembled near it, and, fearless of danger, picks up the scattered crumbs and seeds. His voice is not heard in the spring so early as that of the Song-Sparrow and the Bluebird. He lives chiefly upon seeds, though like other granivorous birds he feeds his young with larvæ. This is a general practice among the seed-eaters, in order to provide their young with soft and digestible food. Nature has provided in a different manner, however, for the Pigeon tribe. The parent bird softens the food in its own crop before it is given to the offspring. From the peculiar manner in which the young are fed comes the expression "sucking doves."

It is common to speak disparagingly of the Hair-Bird, as if he were good for nothing, without beauty and without song. He is despised even by epicures, because his weight of flesh is not worth a charge of powder and shot. Though he is contemptuously styled the "Chipping-Sparrow," on account of his shrill note, this name I shall never consent to apply to him. His voice is no mean accompaniment to the general chorus which may be heard on every still morning before sunrise during May and June. His continued trilling note is to this warbling band like the octave flute, as heard in a grand concert of artificial instruments. The voices of numbers of his species, which are the first to be heard and the last to become silent in the morning, serve to fill up the pauses in this sylvan anthem like a running accompaniment in certain musical compositions. How little soever the Hair-Bird may be valued as a songster, his voice, I am sure, would be most sadly missed, were it nevermore to be heard charmingly blending with the louder voices of other choristers.

How often, on still sultry nights in summer, when hardly a breeze was stirring, and when the humming of the moth might be plainly heard as it glided by my open window, have I been charmed by the note of this little bird, uttered trillingly from the branch of a neighboring tree. He seems to be the sentinel whom Nature has appointed to watch for the first gleam of dawn, which he always faithfully announces before any other bird is awake. Two or three strains from his octave pipe are the signal for a general awakening of the birds, and one by one they join the song, until the whole air resounds with an harmonious medley of voices.

The Hair-Bird has a singular habit of sitting on the ground while thus chirping at early dawn; but I am confident he is perched in a tree during the night. The nest is most frequently placed upon an apple-tree, or upon some tall bush, seldom more than ten feet from the ground. I have found it in the vinery upon the trunk of an elm. It is very neatly constructed of the fibres of roots firmly woven together, and beautifully lined with fine soft hair, whence his name. It is unsurpassed in neatness and beauty by the nest of any other bird. The eggs are four in number, of a pale blue with dark spots.

## THE AMERICAN GOLDFINCH.

During all the pleasant days of autumn, when the thistle and sunflower are ripening their seeds, after the songs of the birds have ceased, and we greet them only as friends after the concert is over, we hear the plaintive chirping of the Hemp-Birds, and see the frequent flashing of their golden plumage among the thistles and goldenrods. Like butterflies they are seen in all the open pastures and meadows that abound in compound flowers, not in flocks, but scattered in great numbers, and always, when flying from one field to another, uttering their singularly plaintive but cheerful cry. This is so sweetly modulated that, when many of them are assembled, the songs of early summer seem to be temporarily revived. They are very familiar and active, always flitting about our flower-gardens when they abound in marigolds and asters.

The Hemp-Bird bears considerable resemblance to the Canary in his habits and the notes of his song. Being deficient in compass and variety, he cannot be ranked with the finest of our songsters. But he has great sweetness of tone, and is equalled by few birds in the rapidity of his execution. His note of complaint is also like that of the Canary, and is heard at almost all times of the year. He utters, when flying, a rapid series of notes during the repeated undulations of his flight, and they seem to be uttered with each effort he makes to rise.

The female does not build her nest before the first broods of the Robin and the Song-Sparrow have flown. Mr. Augustus Fowler, of Danvers, thinks, from his observation of the habits of these birds when feeding their young, that the cause of this delay is "that they would be unable to find in the spring those milky seeds which are the necessary food for their young," and takes occasion to allude to that beneficent law of Nature providing that these birds "should not bring forth their young until the time when the seeds used by them for food have passed into the milk, and may be easily dissolved by the stomach."

These little birds are remarkable for associating at a certain season, and singing as it were in choirs. " During spring and summer," says Mr. Fowler, "they rove about in small flocks, and in July will assemble together in considerable numbers on a particular tree, seemingly for no other purpose than to sing. These concerts are held by them on the forenoon of each day for a week or ten days, after which they soon build their nests. I am inclined to believe that this is the time of their courtship, and that they have a purpose in their meetings beside that of singing. If perchance one is heard in the air, the males utter their call-note with great emphasis, particularly if the new-comer be a female; and while, in her undulating flight, she describes a circle preparatory to alighting, they will stand almost erect, move their heads to the right and left, and burst simultaneously into song."

While engaged in these concerts it would seem as if they were governed by some rule that enabled them to time their voices, and to swell or diminish the volume of sound. Some of this effect is undoubtedly produced by the gradual manner in which the different voices join in harmony, beginning with one or two and increasing their numbers in rapid succession, until all are singing at once, and then in the same gradual manner becoming silent. One voice leads on another, the numbers multiplying, until they make a loud shout which dies away gradually, and a single voice winds up the chorus. These concerts are repeated at intervals for several days, ending probably with the period of courtship.

A singular habit of the Hemp-Bird is that of building a nest, and then tearing it to pieces, before any eggs have been laid in it, and using the materials to make a new nest in another place. When I was a student I repeatedly observed this operation in some Lombardy poplars that grew before my study windows. I thought the male bird only addicted to this habit, and that it might be his method of amusing himself before his mate is ready to occupy the nest. This is made of cotton, the down of the fern, and other soft materials woven together with threads or the fibres of bark, and lined with cow's-hair. It is commonly placed in the fork of the slender branches of a maple, linden, or poplar, and is fastened to them with singular ingenuity.

### THE PURPLE FINCH OR AMERICAN LINNET.

The American Linnet is almost a new acquaintance of many people in Eastern Massachusetts. In my early days, which were passed in Essex County, I seldom met one in my rambles. It is now very common in this region, and has been more generally observed since the custom of planting the spruce and the fir in our gardens and enclosures. The Linnet, though not early in building its nest, is sometimes heard to sing earlier even than the Song-Sparrow. I have frequently heard his notes in March; and once, in a mild season, I heard one warbling cheerily on the 18th of February. But the Linnet does not persevere like the Song-Sparrow and other early birds. He may sing on a fine day in March, and you may not hear him again before the middle of April. Soon after that time he becomes a very constant singer.

The notes of this bird are very simple and melodious, delivered without precision, and different individuals differ exceedingly in capacity. It is generally believed that the young males are the best singers, and that age diminishes their vocal powers. This is the supposition of Mr.

Nuttall; but I have not been able to test the truth of it by my own observation. The greater number utter only a few strains, resembling the notes of the Brigadier. These are constantly repeated during the greater part of the day. The song usually consists of four or five strains, very much alike; but when the bird is animated he multiplies his notes *ad libitum*, varying the modulation only by greater emphasis. I have not observed that the Linnet is more prone to sing in the morning and evening than at any other hour.

The Linnet is a somewhat eccentric bird in his ways. He is usually high up in an elm or other tall tree when he sings, and almost out of sight, like the Brigadier. Hence he is as often heard in the elms in the city as in the country. He sings according to no rules, at no particular hour of the day, with but little regard to season, and utters notes that are wholly wanting in precision. His song is without a theme, and seems to be a sort of *fantasia*. He may often be seen sitting on a fence warbling with ecstasy and keeping his wings in rapid vibration all the while. He is also regardless of the mischief he may do. He feeds upon the flower-buds of the elm and then upon those of the pear-tree, thus damaging our gardens and keeping himself at a safe distance from the angry horticulturist after he has concluded his feast.

I have seen the Linnet frequently in confinement, which he very cheerfully bears; but he will not sing if he be placed near a Canary-Bird, nor does he at any time sing so well as in a state of freedom. He likewise changes his plumage; and soon, instead of a little brown bird with crimson neck, you see one variously mottled with brown and buff. The finest and most prolonged strains are delivered by the Linnet while on the wing. On such occasions only does he sing with fervor. While perched on a tree his song is usually short and not greatly varied. I think there may be less difference than is commonly supposed in the powers of individuals, and that the songs of the same warbler vary with his feelings. If you closely watch one on a tree while singing, he may be observed suddenly to take flight, and while poising himself in the air, though still advancing, to pour out a continued strain of melody with all the rapture of a Skylark.

The male American Linnet is crimson on the head, neck, and throat, dusky on the upper parts of his body, and beneath somewhat straw-colored. It is remarkable that some of the males are wanting in the crimson head and neck, being plainly clad, like the female. These are supposed to be old birds, and the loss of color is attributed to age. I am doubtful of this, for it can hardly be supposed that any bird can escape the gunner long enough to become gray with age. The only nests of this bird which I have seen were upon spruce-trees. The eggs are of a pale green with dark spots of irregular size.

### THE PEABODY-BIRD.

In the northern parts of New England only are the inhabitants familiar with the habits of the Peabody-Bird, or White-throated Sparrow. I have seen it, however, in Cambridge; and during a season when the currant-worm was very destructive, one individual came frequently into my garden and employed himself in picking the caterpillars from a row of currant-bushes. As the fruit was then ripened, or partially ripe, his appearance so late in the season led me to infer that he had probably a nest somewhere in the Cambridge woods. This is a large Sparrow, and a very fine singing-bird. Samuels says: "The song of this species is very beautiful. It is difficult of description, but resembles nearly the syllables 'chea, dêe de ; dê-d-de, dê-d-de, dê-d-de, dê-d-de, uttered first loud and clear, and rapidly falling in tone and decreasing in volume. This is chanted during the morning and the latter part of the day. I have often heard it at different hours of the night, when I have been encamped in the deep forest, and the effect at that time was indescribably sweet and plaintive. The fact that the bird sings often in the night has given it the name of the Nightingale in many places, and the title is well earned."

The inhabitants of Maine mention this bird as singing late in the season. This is caused by his delay in building his nest, which is not done before June. The words used by the Peabody-Bird in his song are thus described in that State: —

All day whittling, whittling, whittling, whittling.

# THE EARLY FLOWERS.

AMONG the vernal flowers are usually classed all those which in propitious seasons are open during the month of April, like the ground-laurel, the draba verna, and the hepatica, also during the month of May, like the anemones, violets, bellworts, and Solomon's seals, which are among the true Mayflowers. Within the space of these two months the most delicate and interesting flowers of the whole year come to perfection, beginning with the epigæa and hepatica, and bringing along in their rear myriads of bellworts, ginsengs, anemones, saxifrages, and columbines, until the procession is closed by the cranesbill, that leads forth the brilliant host of summer.

The vernal flowers are mostly herbaceous and minute. They grow in sheltered situations on the southern slopes of declivities and the sunny borders of a wood, and require but a short period of heat and sunshine to perfect their blossoms. They are generally pale in their tints, many of them white, and often tinged with delicate shades of blue or lilac. The anemones of our woods are our true Mayflowers. They seldom appear before the first of May, and there is hardly a solitary one to be seen after the first week in June. The ground-laurel, vulgarly called Mayflower, is usually in perfection in the middle of April, and, except very far north, is out of bloom by the middle of May. There are some of our early flowers that remain in perfection during a part of the summer, until they lose their charms by constantly intruding themselves upon our notice. Such are the common buttercups, which are favorites of children when they first appear, but shine like gilded toys, and symbolize no charming sentiment to endear them to our sight.

One of the earliest flowers of April, appearing about two weeks later than the ground-laurel, on the sunny slope of a hill that is protected by woods, and continuing to put forth its delicate blossoms during about five weeks from its first appearance, is the hepatica, or liverwort. They are the flowers that have generally rewarded my earliest botanical rambles, and every year I behold them with increased delight. They are often seen in crowded clusters, half concealed by dry oak-leaves, that were elevated by the flowers as they developed their petals. They vary in color from purple or lilac to lighter shades of the same tints. Appearing in heads that often contain more than twenty flowers, they form a pleasing contrast with the little wood anemones that spangle the mossy knolls with their solitary drooping blossoms. The rue-leaved anemone differs from both of these. More lively in its appearance than either, it bears several upright flowers upon one stalk, with such a look of animation that they seem to smile upon us from their green, shady nooks.

Not the least charming of our Mayflowers is the houstonia, which has no English name that has become popular. As early as the middle of May its flowers are often so thickly strewn over the fields as at a distance to resemble a thin veil of snow. This plant is almost as delicate as the finer mosses, and its flowers, though minute, are rendered conspicuous by the brilliant golden hue of their centre, that melts into the cerulean whiteness of the corolla. About the first of May a few flowers of this species peep out from the ground, as in early evening a few stars are seen twinkling through the diminishing light. They multiply until they glitter in the meads and valleys like the heavenly hosts at midnight. By degrees they slowly disappear until June scatters them from the face of the earth, as morning disperses the starry lights of the firmament. It may seem remarkable that the earliest flowers that come up under a frosty sky, and are often enveloped in snow, should, notwithstanding this apparently hardening exposure, exceed all others in delicacy. Such are the ground-laurel, the anemone, and the houstonia, among our native plants, and the snowdrop, the crocus, and the hyacinth, among exotics.

Children, who are unaffected lovers of flowers, have always shown a preference for those of early spring, when they are more attractive on account of their novelty, and seem more beautiful as the harbingers of a warmer season. After the earth has remained bleak and desolate for half the year, every beautiful thing in nature has a renewed charm when it reappears, and a single violet by the wayside inspires a little child with more delight than he would feel if surrounded by a whole garden of flowers in summer.

Parties of young children are annually called out by the first warm sunshine in May to hunt for early flowers. The botanist is also out among the birds and children, peeping into green dells, under shelving rocks, or in sunny nooks, brushing away the dry autumn leaves to find the pale blue liverwort, dipping his hands into crystal streams for aquatic plants, or examining the drooping branches of the andromeda for its rows of pearly gems. He thinks not meanly of his pursuit, though he finds for his companions the village children, and the poor herbwoman, who is gathering salads for the market. From her lips he may obtain some important knowledge, and derive a moral hint that the sum of our enjoyments is proportional to the simplicity of our habits and pursuits, and that this poor herbwoman, who lives chiefly under the open windows of heaven, enjoys more happiness than many envied persons who are prisoned in a palace and shackled with gold. By talking with the children he may learn the locality of some rare plant, a new phase in the aspect of nature, or discover some forgotten charm that once hovered round certain old familiar scenes to whose cheering influence he had become blunted, but which is now revived by witnessing its effects on the susceptible minds of the young.

We have to lament in this climate the absence of many beautiful flowers which are associated in our minds with the opening of spring by our familiarity with English literature. We search in vain over our green meads and sunny hillsides for the daisy and the cowslip, that spangle the fields in Great Britain and gladden the sight of the English cottager. We have read of them until they seem like the true tenants of our own fields; and when on a pleasant ramble we do not find them, there is a void in the landscape, and the fields seem to be wanting in their fairest ornaments. Thus poetry, while it inspires the mind with sentiments that increase the sum of our happiness, often binds our affections to objects we can never behold and shall never caress. The daisy and cowslip are remembered in our reading as the bright-eyed children of Spring, and they emblemize those little members of our former family circle of whom we have heard but have never seen, who exist only in the pensive history of the youthful group whose number is imperfect without them.

In our gardens only do we find the pensive snowdrop, the poetic narcissus, the crocus, and the hyacinth. There only is the pansy, or tri-colored violet, which adorns the fresh chaplets of April and blends its colors with the yellow sheaves of autumn. There only are the lily of the valley, the white Bethlehem star, and the blue-eyed periwinkle. The heath is neither in our fields nor in our gardens. The flowers of classic lands and many plants which are sacred to the muse are not in the fields and valleys of the new continent. Our native flowers, for the most part, are rendered sacred only by the recollections of childhood, not by poetry or romance. The anemone, the houstonia, and the bellwort look up to us from their mossy beds full of the light of the happy days of our youth; but the flowers which have been sung by the Grecian or Roman muse belong to other climes, and our fields do not know them.

# ROCKS.

It is not necessary that an object should be intrinsically beautiful, like a collection of water, to add a pleasing feature to the landscape. Though rocks considered apart from Nature are unsightly, no scenery is complete without them. To a prospect they afford variety which it would be difficult to obtain from any other objects. Without them there is a want of those sudden transitions from the smooth to the rough, from the level to the precipitous, from the beautiful to the wild, and from the tame to the expressive, which are essential to a perfect landscape. It is only among rocks that the evergreen ferns, those beautiful accompaniments of a rustic retreat, are found growing abundantly. There is no more beautiful sight than a series of almost perpendicular rocks covered on all sides by ferns, with their peculiarly graceful foliage, and here and there a rill trickling down their surface and forming channels through the evergreen mosses. The solitary glens formed by these rocks could not be imitated by any artifice; and their jutting precipices afford prospects unequalled by the gentle elevations of a rolling landscape. In a country where rocks are wanting, the land rises and sinks in gradual declivities, and prospects are difficult to be obtained except from lofty elevations.

There is so much that is attractive in the abruptness of rocky scenery, especially when half covered by trees and other vegetation, that some authors have attributed its picturesque character to its rudeness and roughness. I should attribute this interesting expression to the manifest facility which abrupt situations afford both for prospect and for pleasant secluded retreats. Large clefts produced by the parting of the two sides of an enormous rock furnish dells, — often perfect gardens of wild-flowers, — bursting on the sight like an oasis on a dry waste. In these places there is always a remarkable verdure, as the rains that pour down the slopes conduct fertility to the soil at their base. A rocky surface, therefore, is productive of a greater variety of shrubs and wildflowers than a plain or rolling country of similar soil and climate.

There are many plants whose native localities are the tops and sides of rocky cliffs and precipices. Such are the saxifrage, the cistus, the toadflax, and the beautiful pedate violet. The graceful Canadian columbine is found chiefly among the clefts of rocks, like a little tender animal, nestling under their protection, and drawing nourishment from the soil that has accumulated in their hollows. To satisfy ourselves of the number and variety of plants that may grow spontaneously upon a single rock, let us construct one in fancy thus enamelled by the hand of Nature.

We will picture to ourselves a craggy precipice, rising thirty or forty feet out of a wet meadow, and forming in its irregular ascent many oblique and perpendicular sides, which have collected upon their upper surface several inches of soil. A grove of pines and birches covers the summit, with an undergrowth of various shrubs, such as the whortleberry, the wood-pyrus, the spiræa, and the mountain andromeda. Here, too, the bayberry and sweet fern mingle their fragrance with the odors of the pines. The rocks, in the driest places, are covered with a bedding of gray lichen, which is a perfect hygrometer, breaking like glass under our footsteps when the atmosphere is dry, but yielding like velvet when it contains the least moisture. The cup-moss grows abundantly along with it, and in moist situations the green, delicate hairmoss, the same that covers the roofs of very old buildings. The rain has washed down from the summit constant deposits from trees and shrubs, birds and quadrupeds, and formed a superficies of good soil on all parts of the rock where it could be retained. On the almost bare surface grows the beautiful feather-grass, supported only by the soil that has accumulated about its roots.

The mountain-laurel luxuriates upon these natural terraces, by which we descend to the meadow at the base of the rock. But this evergreen, with its magnificent clusters of flowers, is not the most attractive object, for the little springs that issue from the crevices of the rock have brought out a variety of ferns and lycopodiums that cover its sides with their green fronds, like the tiles on the roof of a house. Some oaks and beeches project fantastically from the sides of the cliff, which is covered with innumerable vines. Beside the beautiful things that cluster at our feet, and the little winged inhabitants native to the situation, made attractive by their various forms, colors, and motions, this rock gives additional extent to the prospect of the surrounding country, and affords many different views from the various openings through its wood and shrubbery.

Such are the beauties and advantages multiplied about a mere rock. But in my description I have omitted to notice the grotto formed by the shelving of rocks, so delightful to the traveller who seeks shelter from the sultry heat of noon, or to one who only wishes to gratify a poetic sentiment. Rocky scenery cannot fail to suggest to the mind the various scenes and incidents of romantic adventure; and I believe the difficulties and dangers it presents to the traveller magnify the interest attending it. I have often seen a whole party eager to obtain possession of a flower that was growing out of the edge of a rocky cliff. Each one would feel a desire to climb upon its sides, and to obtain a resting-place upon its dangerous summit.. These circumstances stimulate the adventurous spirit, and become picturesque when represented on canvas, by affording the same agreeable excitement to the imagination. Hence the imaginative as well as the adventurous are delighted with this kind of scenery, that arouses the enterprise of the one and awakens the poetical feelings of the other. What do we care for a scene, however beautiful, which is so tame as to offer no exercise for the imagination? Rocks, by increasing the inequalities of the surface, proportionally multiply the ideas and images that are associated with a landscape.

It is not an uninteresting inquiry, why a prospect beheld from a rocky cliff yields us more pleasure than the same beheld from an even slope. Is it more poetical, when we partake of any such enjoyment, to be disconnected from objects immediately around us? Or, when standing upon a rock that projects from the surface of the ground, may we not experience an illusive feeling of elevation? On the northern coast of Massachusetts Bay are many grand and delightful views of the ocean from points on the neighboring hills and eminences. Some of these views are unsurpassed in beauty. I have repeatedly observed that parties of pleasure, when making an excursion on the hills, are not satisfied with a view of the sea and the landscape until they have beheld it from some towering rock. There is probably a poetic feeling of isolation attending us when standing upon a rock that increases the emotions, whether of beauty or sublimity, which are excited by the prospect.

Any one who has rambled over the bald hills that bound this shore can bear witness to the power of such rude scenery to magnify the sentiments that spring from the aspect of desolation. They are felt in these places, unaccompanied by the melancholy that attends us on surveying a wide scene of ruins. Here the appearance of desolation is sufficient to produce a sentiment of grandeur; but while surrounded by the evidences appearing in a distant view of a fertile and prosperous country, we are equally affected with a sense of cheerful exaltation. The most beautiful garden that wealth and taste could design would not afford so much of the luxury of sentiment as a ramble over these bald hills affords to one whose mind is rightly attuned for such enjoyments. It is evident that the hills without the rocks would be destitute of one of their most charming features. From the sight of the rocks comes likewise that feeling of alliance with the past ages of the world which tends greatly to elevate the mind with sentiments of grandeur.

The New England stone-wall, as a feature in landscape scenery, is generally considered a deformity; yet it cannot be denied that the same lines of wooden fence would mar the beauty of our prospect in a still greater degree. On account of the loose manner in which the stones are laid one upon another, as well as the character of the materials, this wall harmonizes with the rude aspects of nature better than any kind of masonry. It seems to me less of a deformity than a trimmed hedge or any other kind of a fence, except in ornamented grounds, of which I do not treat. In wild pastures and lands devoted to common rustic labor, the stone-wall is the most picturesque boundary-mark that has yet been invented. A trimmed hedge in such places would present to the eye an intolerable formality.

One of the charms of the loose stone-wall is the manifest ease with which it may be overleaped. It menaces no infringement upon our liberty. When we look abroad upon the face of a country subdivided only by long lines of loose stones, and overgrown by vines and shrubbery, we feel no sense of constraint. The whole boundless prospect is ours. An appearance that cherishes this feeling of liberty is essential to the beauty of landscape; for no man can thoroughly enjoy a scene from which he is excluded. Fences are deformities of prospect which we are obliged to use and tolerate. But the loose stone-wall only is expressive of that freedom which is grateful to the traveller and the rambler.

It may be remarked that no inconsiderable share of the interest added to a prospect by the presence of rocks arises from their connection with the past ages of the world. They are indeed the monuments of the antediluvian period; and no man who is acquainted with the most commonly received geological facts, when wandering among these relics of the mysterious past, can fail to be inspired with those emotions of sublimity that proceed less from the creations of poetry than from the wonders of science.

# MARCH.

To the inhabitants of a variable climate, like our own, the weather is at all times one of the most interesting themes of speculation; but at no period of the year does it come more directly home to our feelings than in March. We know that there is a new sign in the heavens, and the altitude of the sun in his meridian seems plainly to assure us of the comforts of spring. But the aspect of the heavens is constantly changing, the winds ever veering, clouds alternating with sunshine, wind with calm, and rain with snow; so that we are never sure, on a bland morning in March, when the sun is shining almost with the fervor of summer, that we may not be overtaken by a snow-storm before noonday, or the cold of the Arctic Circle before sunset. Any one of the three winter months, though usually cold and stormy, may once in a few years be mild and pleasant from beginning to end; but March preserves the same variable and boisterous character from year to year, without deviating from its precedents. It is the only month when day's harbingers never fulfil their promises, — when the rosy hours that come up with the morning and the fair sisters that weave the garlands of evening are all deceivers.

Though the present time is nominally the spring of the year, there is not yet a flower in the fields or gardens, and the buds of the trees are hardly swollen with waking vegetation. The wild-flowers are still buried under the snows and ices of winter, and the grass has begun to look green only under the southern protection of the walls and fences. Many of the early birds, following the southerly winds that often prevail for a few days, and tempted by the bright sunshine of the season, have arrived from their winter haunts, and sing and chirp alternately, as if they were debating whether to remain here or return to a more genial clime. It is a remarkable instinct that prompts so many species of birds to leave their pleasant abiding-places at the south, where every agreeable condition of climate, shelter, and provision for their wants is present, and press onward into the northern regions, before the rigors of winter have been subdued, and while they are still liable to perish with cold or starvation. Often with anxiety have I watched these little bewildered songsters who have so unseasonably returned, when, after commencing their morning lays as if they believed the vernal promises of dawn, they were obliged to flee into the depths of the woods to find shelter from a driving snow-storm.

It may seem remarkable that before vegetation has awakened there should be a revival of some of the insect tribes; but in warm, sheltered situations many small flies may be seen, either newly hatched or revived by the heat of the sun. They do not seek food, but crawl about in dry places, sometimes rising into the air and drowsily and awkwardly exercising their wings. So exposed are these minute creatures to the mercy of the climate, that Nature has made them insusceptible of injury from the severest cold. Many species, though enclosed in a mass of solid ice, may be revived by gradual heat and fly abroad as gayly as if they had been refreshed by sleep. But the period of life assigned to the insect race is very short, and before the arrival of winter the brief and joyous existence of nearly all the species is terminated, and their offspring in an embryo state lie torpid until a new spring awakes them into life.

Our climate, being a discordant mixture of the weather of two opposite latitudes pouring their winds alternately upon our territory, is the most variable and deceitful in the world. Alternating with each other and struggling as it were for the mastery are two winds, — one that sweeps across the Canadas and brings with it the cold of the polar regions, another that comes from the Gulf of Mexico and brings here the summer breezes of the tropics. No natural barrier is interposed to check their progress whenever any meteoric influence may urge them onward. The prevalence of a moderate temperature in this part of the country during a calm, either in spring or autumn, proves this to be the true weather of our latitude. The north and south winds are intruders that spoil the comfort we might otherwise enjoy in the open air at all seasons except the three months of winter. Our climate may, therefore, not unaptly be compared to a village that is peopled by quiet and peaceable inhabitants, but visited by troublesome people from the adjoining villages, who by their quarrels with each other keep it in a constant uproar, leaving the villagers only an occasional respite during their absence.

March is persistent only in its variableness. If it be cold, heat will soon succeed; if we have clouds, they will soon bring along a clear sky. We see none of those melancholy clouds, so common in the latter part of autumn, that remain for weeks brooding over the landscape, as if the heavens were hung in mourning for the departure of summer, — none of that ominous darkness in the glens and valleys, denoting that the sun has at length surrendered to the frosty conqueror of the earth. Though March is colder, it has more light than November. The sun daily increases in power, and the snow that remains upon the earth renders the effect of his rays more brilliant and animating. The clouds of this month are seldom motionless. They are borne along rapidly by the brisk winds, now enveloping the landscape in gloom, then suddenly illuminating it with sunshine, and producing that constant play of light and shade which is peculiar to the early spring.

During the occasional days of pleasant serenity that occur in March, we begin to look about us among the sheltered retreats in the woods and mountains, to watch the earliest budding of vegetation. Seldom, however, do we find a flower outside of the gardens; but many a green herb, that has been preserved under the snow or under the protection of shrubbery, may be seen creeping upon the surface, and spreading its delicate verdure upon the brown turfs. There the leaves of the strawberry and the cinqfoil are as green as in summer, and the tall hypericum, which is as it were a tree in summer, becomes in winter and spring a creeping vine, with foliage as fine as that of a heath. At such times, while sauntering about the fields, rejoicing in what seems to be a true revival of spring, the fierce north-wind begins his raging anew, and ere another morning the birds lie concealed in the depths of the forest, and all hearts are saddened by the universal aspect of winter.

The change that has taken place in the appearance of the sun at his rising, since the opening of this month, may be regarded as one of the usual indications of the reviving spring. The atmosphere, on clear mornings, is more heavily laden with vapors than is usual at the same hour in winter. The exhalations of the preceding day have been descending in frosty dews by night upon the plains, and while gathered thickly about the horizon yield to the first rays of the sun a tint of purple and violet, like the dawn of a summer morning. The sun in midwinter, when there are no vapors on the lakes and meadows, after the cold winds have frozen every source of exhalant moisture, rises into a clear, transparent atmosphere. As spring advances, and the sun rises higher, the evaporation increases, the atmosphere in the morning becomes charged with prismatic vapors, and every mead and valley is crowned at sunrise with wreaths of mist adorned with the hues of the rainbow. The crimson haze that accompanies the dawn denotes that the icy fountains are unlocked, and that the lakes and rivulets are again pouring their dewy offerings to the skies.

March is an unpleasant month for rambling. There is but little to tempt the lover of Nature, in either field or wood, to examine her treasures, or to enjoy the luxury of climate; but there is still a motive for roaming abroad, though it were but to watch the breaking up of the ice, and to mark the progress of the thousands of new-born rivulets that leap down the snowy mountains toward the grand reservoir of waters. There are places always to be found which are inviting to the solitary pedestrian during the most uncomfortable seasons. The fairy hands that were once busy in spreading tints upon the flowers and upon the heavens still toil unseen in their deserted places, weaving the few fragments of remaining beauty upon moss-grown hillocks and in fern-embroidered nooks.

People who have always lived in the interior of the country can have only a feeble conception of the pleasure of a seaside ramble, which is during this month, when the west-winds prevail the greater part of the time, more agreeable than a walk in the open plain. Among the lakes and rivers and hills and valleys of an interior landscape, though there be an endless variety of pastoral beauty, there is nothing that will compare with the grandeur of a water prospect from the sea-shore. Neither can such a view be fully appreciated by those who have beheld it only from the harbor of a large city, where the works of art cover and conceal its native magnificence, and withdraw the mind from those poetic thoughts that would be awakened by an unsophisticated ocean-scene. We must go forth upon the solitary shores, at a distance from all artificial objects, and walk upon the high bluffs that project far enough into the sea to afford sight of a complete hemisphere of waters, to obtain a just idea of a sea-prospect. When we look from the deck of a sailing ship, where nothing on any side is to be seen but the ocean, bounded by the circle that seems to divide the dark blue of the waters from the more ethereal azure of the skies, our sublime emotions are not modified by any sensations of beauty; but when this blue expanse of waters divides the prospect equally with the landscape that is spread out in a luxuriant variety of wood, plain, and mountain, the emotions excited by the sublimity of the scene are softened into repose by the beauty and loveliness of the opposite prospect.

But the sun is daily rising higher into the zenith. The blustering winds are losing their force and are yielding to the fate that awaits them inevitably after the winter has passed away. The trees bow their heads less lowly to the gales, standing more and more erect, as if conscious that the time of their triumph is near, and that the singing-birds are awaiting the opening of their flowers and the unfolding of their leaves. The infant Spring is fast becoming a maiden and a goddess, and the herbs are preparing to weave garlands for her virgin brows, daisies to spread at her feet, and ambrosial incense, such as in heaven surrounds the presence of purity and holiness, to gladden her coming. Let the winds rage, and the clouds threaten; we know that soon their anger will be quelled by the genial sunshine of spring, as the tumults in the human breast are tranquillized by the smiles of innocence and beauty.

# SINGING-BIRDS.

THE Singing-Birds, with reference to their songs, are distinguishable into four classes: — The Rapid singers, whose song is uninterrupted, of considerable length, and delivered in apparent ecstasy; the Moderate singers, whose notes are slowly modulated, without pauses or rests between the different strains; the Interrupted singers, who sometimes modulate their notes with rapidity, but make a distinct pause after each strain. The Linnet and the Bobolink are examples of the first class; the common Robin and the Veery of the second; the Red Thrush and particularly the Hermit Thrush of the third. There are other birds whose lay consists only of two or three notes, not sufficient to be called a song. The Bluebird and the Golden Robin are of this class.

June, in this part of the world, is the most tuneful month of the year. Many of our principal songsters do not appear until near the middle of May; but all, whether early or late, continue to sing throughout the month of June. The birds that arrive the latest are not always the latest in returning. The period of time they occupy in song depends chiefly upon the number of broods of young they raise in the year. If they raise but one brood in a season, their period of song is short; if they raise two or more, they may prolong their singing into August. Not one of our New England birds is an autumnal warbler, though the Robin, the Wood-Sparrow, and the Song-Sparrow are often heard after the first of September. The tuneful season in New England comprises April, May, and the three summer months.

There are certain times of the day, as well as certain seasons of the year, when birds are most musical. The grand concert of the feathered tribe takes place during the hour between dawn and sunrise. During the remainder of the day until evening they have no concerts. Each individual sings according to its habits, but we do not hear them collectively. At sunset there is an apparent attempt to unite once more in chorus, but this is far from being so loud or so general as in the morning, when they suffer less disturbance from man.

There are but few birds whose notes could be accurately described upon the gamut. We seldom perceive anything like artificial pauses or true musical intervals in their time or melody. Yet they have no deficiency of musical ear, for almost any singing-bird when young may be taught to warble an artificial tune. Birds do not dwell steadily upon one note at any time. They are constantly sliding and quavering,. and their songs are full of pointed notes. There are some species whose lays, like those of the Whippoorwill, resemble an artificial modulation, but these are rare. In general their musical intervals cannot be accurately distinguished on account of the rapidity of their utterance. I have often endeavored to transcribe their notes upon the gamut, but have not yet been able to communicate to any person by this means a correct idea of the song, except in 'a few extraordinary cases. Such attempts are almost useless.

Different individuals of certain species often sing very unlike each other; but if we listen attentively to a number of them, we shall detect in all their songs a *theme*, as it is termed by musicians, of which they severally warble their respective variations. Every song of any species is, technically speaking, a fantasia constructed upon this theme, from which, though they may greatly vary their notes, no individual ever departs. The theme of the Song-Sparrow is easily written on the gamut, out of which the bird makes many variations; that of the Robin's song is never more than slightly varied; but I have not been able to detect in the medley of the Bobolink any theme at all.

The song of birds is innate. It is not learned, as some have supposed, from parental instruction; else why should not a Cowbird sing like a Vireo, which is sometimes its foster parent, and would undoubtedly, if this were the usual custom, be as willing to teach the young interloper to sing as to supply it with food? Birds of the same species have by their organization a disposition to utter certain sounds when under the influence of certain feelings. If the young bird learned of its parents, nature would have made the female the singer instead of the male, who, I am confident, would not trouble himself to be a music-teacher, and, if he were willing to take this task upon him, would not select the males only to be his pupils. If we should see repeated instances of the exemplification of their mode of instruction, — if we should see the young birds standing around an old cock Robin while he delivers his song, note by note, for the young to imitate, — we should have some reason to believe that all male singing-birds are music-teachers as well as performers. But after all, would an old Bobolink ever have patience to repeat his notes slowly to his young for their instruction?

Many birds are, however, imitators of sounds, and will sometimes learn the songs of other birds when confined in a cage near them. The Bobolink when caged near a Canary readily learns its song, but in a wild state he never deviates from his own peculiar medley. Nature has provided each species with notes unlike those of any other as one of the means by which they should identify their own kindred, and there is reason to believe that if one of them had never heard the note of his own parents he would still sing like all his predecessors. In a state of confinement birds will occasionally imitate the notes of other species, and in this respect they differ entirely from quadrupeds.

The song of birds seems to be the means used by the male, not only to woo the female, but to call her to himself when absent. Before he has chosen his mate he sings more loudly than at any subsequent period. The different males of the same species seem at that time to be vying with each other, and the one that has the loudest and most varied song is likely to be the first attended by a mate. When the two birds are employed in building their nest, the male constantly attends his partner and sings less loudly and frequently than before. This comparative silence continues until the female begins to sit. During incubation the male again sings with emphasis at his usual hours, perched upon some neighboring tree, as if to assure her of his presence, but more probably to entice her away from the nest. It is a curious fact that male birds seem to be displeased to a certain extent while their mate is sitting, on account of her absence, and are more than usually vociferous, sometimes with the evident intention of coquetting with other females.

After the young brood is hatched the attention of the male bird is occupied with the care of his offspring, though he is far less assiduous in his parental duties than the female. If we watch a pair of Robins when they have a nest full of young birds, we shall see the female bring the greater part of their food. The male bird continues to sing until the young have left their nest; but if there is to be no other brood, he becomes immediately silent. If, early in the season, a couple whose habit is to rear but one brood are robbed of their nest, they will make a new one, and the male in this case continues in song to a later period than those who were not disturbed.

If the male bird loses his mate during incubation, he seldom takes her place, but becomes once more very tuneful, uttering his call-notes loudly for several days and finally changing them into song. It would seem, therefore, that the song of the bird proceeds in some degree from discontent,— from his want of a mate, in the one case, or from her absence when she is sitting, in the other. The buoyancy of spirits produced by the season and the full supply of his physical wants are joined with the pains of absence, which he is determined to relieve by exerting all his power to entice his partner from her nest. I have often thought that the almost uninterrupted song of caged birds proves their singing to arise from a desire to entice a companion into their own little prison. Hence, when an old bird from our fields is caught and caged during the breeding-season, he will continue his tunefulness long after all others of the same species have become silent. The Bobolink in a state of freedom will not sing after the middle of July; but if one be caught and caged, he will continue to warble more loudly than he did in his native meadows until September.

It is generally believed that singing-birds are chiefly confined to temperate latitudes. That this is an error is apparent from the testimony of travellers, who speak of the birds of Africa and of the Sandwich Islands as singing delightfully; and some fine songsters are occasionally imported from tropical countries. It should be considered that in these hot regions the birds are more scattered and are not so well known as those of temperate latitudes, which are generally inhabited by civilized man. Savages and barbarians, who are the principal inhabitants of hot countries, are seldom observant of the songs or habits of birds. A musician of the feathered race, no less than a human singer, must have an appreciating audience or his powers could not be made known to the world. But even with the same audience, the tropical birds would probably be less esteemed than those of equal merit in our latitudes, for amid the stridulous and deafening sounds from insects in warm climates the notes of birds are scarcely audible. Probably, however, the comparative number of singing-birds is greater in the temperate zone, where there are more of those species that build low, and live in the shrubbery, which the singing-birds chiefly frequent. In warm climates the birds are obliged to live in trees, and the vegetation of the surface of the ground will not support the Finches and Buntings, which are the chief singers of the North.

# BIRDS OF THE GARDEN AND ORCHARD.

## II.

### THE VIREO.

IN the elms on Boston Common, and in all the lofty trees of the suburbs, as well as in the country villages, are two little birds whose songs are heard daily and hourly, from the middle of May until the last of summer. They are usually concealed among the highest branches of the trees, so that it is not easy to obtain sight of them. These birds are two of our Warbling Flycatchers, or Vireos; one of which I shall designate as the Brigadier, the other as the Preacher. I give below the song of the Brigadier: —

Brig - a - dier,   Brig - a - dier,   Brigate.

The notes of this little invisible musician are few, simple, and melodious, and, being often repeated, they are very generally known even to those who are unacquainted with the bird. At early dawn, at noon, and at sunset its song is constantly repeated with no very long intervals, resembling, though delivered with more precision, the song of the Linnet or Purple Finch. In my boyhood, when I had no access to a book descriptive of our birds, and very seldom killed one for any purpose, I had learned nearly all the songs that were heard in the garden or wood, without knowing the physical characters of more than one out of three of the songsters; and as I have since studied the markings of birds only by viewing them from the ground as they were perched upon bush or tree, and have never killed or dissected one for this purpose, I cannot describe all the specific or generic characters of our birds. I am well acquainted with two of our Vireos; but I cannot distinguish them from each other except by their notes, which are as familiar to me as the voice of the Robin. I have, therefore, determined to name them according to the style of their songs, leaving it to others to identify the species to which they respectively belong.

The Brigadier, which is the one, I think, described by Nuttall as the Warbling Vireo, is a little olive-colored bird, that occupies the lofty tree-tops while singing and hunting his food, and is almost invisible as he is flitting among the branches, and never still. The Preacher (Red-eyed Vireo) arrives about a week or ten days earlier than the Brigadier, and is later in his departure. The two are very similar, both in their looks and their habits, frequenting the trees in the town and its suburbs in preference to the woods, singing at all hours of the day, particularly at noon, and taking their insect prey from the leaves and branches of the trees, or seizing it as it flits by their perch, and amusing themselves while thus employed with their oft-repeated notes. Each species builds a pensile nest, or places it in a fork of the slender branches of a tree. I have seen a nest of the Brigadier about ten feet from the ground on a branch of a pear-tree, so near my chamber-window that I might have reached it without difficulty. The usual habit of either species is to suspend its nest at a very considerable height from the ground.

## THE PREACHER.

The Preacher is more generally known by his note, because he is incessant in his song, and particularly vocal during the heat of our long summer days, when only a few birds are singing. His style of preaching is not declamation. Though constantly talking, he takes the part of a deliberative orator, who explains his subject in a few words and then makes a pause for his hearers to reflect upon it. We might suppose him to be repeating moderately, with a pause between each sentence, "You see it, — you know it, — do you hear me? — do you believe it?" All these strains are delivered with a rising inflection at the close, and with a pause, as if waiting for an answer.

The tones of the Preacher are loud and sharp, hardly melodious, modulated somewhat like those of the Robin, though not so continuous. He is never fervent, rapid, or fluent, but, like a true zealot, he is apt to be tiresome, from the long continuance of his discourse. He pauses frequently in the middle of a strain to seize a moth or a beetle, beginning anew as soon as he has swallowed his morsel. Samuels expresses great admiration for this little bird. "Everywhere in these States," he remarks, "at all hours of the day, from early dawn until evening twilight, his sweet, half-plaintive, half-meditative carol is heard," and he adds, that of all his feathered acquaintances this is his favorite. The prolongation of his singing season until sometimes the last week in August renders him a valuable songster. When nearly all other birds have become silent, the little Preacher still continues his earnest harangue, and is sure of an audience at this late period, when he has but few rivals.

## THE BOBOLINK.

There is not a singing-bird in New England that enjoys the notoriety of the Bobolink. He is like a rare wit in our social or political circles. Everybody is talking about him and quoting his remarks, and all are delighted with his company. He is not without great merits as a songster; but he is well known and admired because he is showy, noisy, and flippant, and sings only in the open field, and frequently while poised on the wing, so that any one who hears can see him and know who is the author of the strains that afford so much delight. He sings also at broad noonday, when everybody is out, and is seldom heard before sunrise, while other birds are joining in the universal chorus. He waits till the sun is up, when many of the early performers have become silent, as if determined to secure a good audience before his own exhibition.

In the grand concert of Nature it is the Bobolink who performs the recitative, which he delivers with the utmost fluency and rapidity, and we must listen carefully not to lose many of his words. He is plainly the merriest of all the feathered creation, almost continually in motion, and singing on the wing apparently in the greatest ecstasy of joy. There is not a plaintive strain in his whole performance. Every sound is as merry as the laugh of a young child, and we cannot listen to him without fancying him engaged in some jocose raillery of his companions. If we suppose him to be making love, we cannot look upon him as very deeply enamored, but rather as highly delighted with his spouse and overflowing with rapturous admiration. His mate is a neatly formed bird, with a mild expression of face, of a modest deportment, and arrayed in the plainest apparel. She seems perfectly satisfied with observing the pomp and display of her partner, and listening to his delightful eloquence of song. If we regard him as an orator, it must be allowed that he is unsurpassed in fluency and rapidity of utterance; if only as a musician, that he is unrivalled in brilliancy of execution.

I cannot look upon him as ever in a very serious humor. He seems to be a lively, jocular little fellow, who is always jesting and bantering; and when half a dozen different individuals are sporting about in the same orchard, I can imagine they might represent the persons dramatized in some comic opera. The birds never remain stationary upon a bough, singing apparently for their own solitary amusement; they are ever in company, passing to and fro, often beginning their song upon the extreme end of an apple-tree bough, then suddenly taking flight and singing the principal part while balancing themselves on the wing. The merriest part of the day with these birds is the later afternoon, during the hour preceding dewfall, before the Robin and the Veery begin their evening hymn. At that hour, assembled in company, they might seem to be practising a cotillon on the wing, each one singing to his own movement as he sallies forth and returns, and nothing can exceed their apparent merriment.

The Bobolink begins his morning song just at sunrise, at the time when the Robin, having sung from earliest daybreak, is near the close of his performance. Nature seems to have provided that the serious parts of her musical entertainment in the morning shall first be heard, and that the lively and comic strains shall follow them. In the evening this order is reversed, and after the comedy is concluded Nature lulls us to repose by the mellow notes of the Vesper-Bird, and the pensive and still more melodious strains of the solitary Thrushes.

In pleasant shining weather the Bobolink seldom flies

without singing, often hovering on the wing over the place where his mate is sitting upon her ground-built nest, and pouring forth his notes with the greatest loudness and fluency. Vain are all the attempts of other birds to imitate his truly original style. The Mocking-Bird is said to give up the attempt in despair, and refuses to sing at all when confined near one in a cage. The Bobolink is not a shy bird during the breeding season; but when the young are reared and gathered in flocks the whole species become very timid. Their food consists entirely of insects during at least all the early part of summer. Hence they are not frequenters of the woods, but of the fields that supply their insect food. They evidently have no liking for solitude. They join with their own kindred, sometimes, during the breeding season, in small companies, and in the latter summer in large flocks. They love the orchard and the mowing-field, and many are the nests which are exposed by the scythe of the haymaker when performing his task early in the season.

### THE O'LINCON FAMILY.

A flock of merry singing-birds were sporting in the grove;
Some were warbling cheerily and some were making love.
There were Bobolincon, Wadolincon, Winterseeble, Conquedle, —
A livelier set were never led by tabor, pipe, or fiddle : —
Crying, "Phew, shew, Wadolincon ; see, see Bobolincon
Down among the tickle-tops, hiding in the buttercups;
I know the saucy chap ; I see his shining cap
Bobbing in the clover there, — see, see, see!"

Up flies Bobolincon, perching on an apple-tree;
Startled by his rival's song, quickened by his raillery.
Soon he spies the rogue afloat, curvetting in the air,
And merrily he turns about and warns him to beware !
"'T is you that would a wooing go, down among the rushes O!
Wait a week, till flowers are cheery ; wait a week, and ere you marry,
Be sure of a house wherein to tarry ;
Wadolink, Whiskodink, Tom Denny, wait, wait, wait!"

Every one 's a funny fellow ; every one 's a little mellow ;
Follow, follow, follow, follow, o'er the hill and in the hollow.
Merrily, merrily there they hie ; now they rise and now they fly ;
They cross and turn, and in and out, and down the middle and wheel about,
With a "Phew, shew, Wadolincon ; listen to me, Bobolincon !
Happy 's the wooing that 's speedily doing, that 's speedily doing,
That 's merry and over with the bloom of the clover ;
Bobolincon, Wadolincon, Winterseeble, follow, follow me !"


O what a happy life they lead, over the hill and in the mead !
How they sing, and how they play ! See, they fly away, away !
Now they gambol o'er the clearing, — off again, and then appearing ;
Poised aloft on quivering wing, now they soar, and now they sing,
"We must all be merry and moving ; we must all be happy and loving ;
For when the midsummer is come, and the grain has ripened its ear,
The haymakers scatter our young, and we mourn for the rest of the year ;
Then, Bobolincon, Wadolincon, Winterseeble, haste, haste away !"

### THE BLUEBIRD.

Not one of our songsters is so intimately associated with the early spring as the Bluebird. Upon his arrival from his winter residence, he never fails to make known his presence by a few melodious notes uttered from some roof or fence in the field or garden. On the earliest morning in April, when we first open our windows to welcome the soft vernal gales, they bear on their wings the sweet strains of the Bluebird. These few notes are associated with all the happy scenes and incidents that attend the opening of the year.

The Bluebird is said to bear a strong resemblance to the English Robin-Redbreast, similar in form and size, having a red breast and short tail-feathers, with only this manifest difference, that one is olive-colored above where the other is blue. But the Bluebird does not equal the Redbreast as a songster. His notes are few and not greatly varied, though sweetly and plaintively modulated and never loud. On account of their want of variety, they do not enchain the listener; but they constitute an important part of the melodies of morn.

The value of the inferior singers in making up a general chorus is not sufficiently appreciated. In a musical composition, as in an anthem or oratorio, though there is a leading part, which is usually the air, that gives character to the whole, yet this leading part would often be a very indifferent piece of melody if performed without its accompaniments; and these alone would seem still more trifling and unimportant. Yet, if the composition be the work of a master, these brief strains and snatches, though apparently insignificant, are intimately connected with the harmony of the piece, and could not be omitted without a serious disparagement of the grand effect. The inferior singing-birds, bearing a similar relation to the whole choir, are indispensable as aids in giving additional effect to the notes of the chief singers.

Though the Robin is the principal musician in the general anthem of morn, his notes would become tiresome if heard without accompaniments. Nature has so arranged the harmony of this chorus, that one part shall assist another; and so exquisitely has she combined all the different voices, that the silence of any one cannot fail to be immediately perceived. The low, mellow warble of the Bluebird seems an echo to the louder voice of the Robin; and the incessant trilling or running accompaniment of the Hair-Bird, the twittering of the Swallow, and the loud, melodious piping of the Oriole, frequent and short, are sounded like the different parts in a band of instruments, and each performer seems to time his part as if by some rule of harmony. Any discordant sound that may occur in this performance never fails to disturb the equanimity of the singers, and some minutes will elapse before they resume their song. It would be difficult to draw a correct comparison between the birds and the various instruments they represent. But if the Robin were described as the clarionet, the Bluebird might be considered the flageolet, frequently but not incessantly interspersing a few mellow strains. The Hair-Bird would be the octave flute, constantly trilling on a high key, and the Golden Robin the bugle, often repeating his loud and brief strain. The analogy, if carried further, might lose force and correctness.

All the notes of the Bluebird — his call-notes, his notes of complaint, his chirp, and his song — are equally plaintive and closely resemble one another. I am not aware that this bird utters a harsh note. His voice, which is one of the earliest to be heard in the spring, is associated with the early flowers and with all pleasant vernal influences. When he first arrives he perches upon the roof of a barn or upon some leafless tree, and delivers his few and frequent notes with evident fervor, as if conscious of the pleasures that await him. These mellow notes are all the sounds he makes for several weeks, seldom chirping or scolding like other birds. His song is discontinued at midsummer, but his plaintive call, consisting of a single note pensively modulated, continues every day until he leaves our fields. This sound is one of the melodies of summer's decline, and reminds us, like the note of the green nocturnal tree-hopper, of the ripened harvest, the fall of the leaf, and of all the joyous festivals and melancholy reminiscences of autumn.

The Bluebird builds his nest in hollow trees and posts, and may be encouraged to breed around our dwellings, by supplying boxes for his accommodation. In whatever vicinity we reside, whether in a recent clearing or the heart of a village, if we set up a bird-house in May, it will certainly be occupied by a Bluebird, unless previously taken by a Wren or a Martin. But there is commonly so great a demand for such accommodations, that it is not unusual to see two or three different species contending for one box.

### THE HOUSE-WREN.

The bird whose notes serve more than any other species to enliven our summer noondays is the common House-Wren. It is said to breed chiefly in the Middle States, but is very common in our New England villages, and as it extends its summer migration to Labrador, it probably breeds in all places north of the Middle States. It is a migratory bird, leaving us early in autumn, and not reappearing until May. It builds in a hollow tree like the Bluebird. A box of any kind, properly made, will answer its purposes. But nothing is better than a grape-jar, prepared by drilling a hole in its side, just large enough for the Wren, and setting it up on a perpendicular branch sawed off and inserted into the mouth of the jar. The bird fills it with sticks before it makes a nest, and the mouth of the jar serves for drainage.

The Wren is one of the most restless of the feathered tribe. He is continually in motion, and even when singing is constantly flitting about and changing his position. We see him in a dozen places as it were at the same moment; now warbling in ecstasy from the roof of a shed, then, with his wings spread and his feathers ruffled, scolding furiously at a Bluebird or a Swallow that has alighted on his box, or driving a Robin from a neighboring cherry-tree. Instantly we observe him running along a stone-wall and diving down and in and out, from one side to the other, through its openings, with all the nimbleness of a squirrel. He is on the ridge of the barn roof, he is peeping into the dove-cote, he is in the garden under the currant-bushes, or chasing a spider under a cabbage-leaf. Again he is on the roof of a shed, warbling vociferously; and these manœuvres and peregrinations have occupied hardly a minute, so rapid and incessant are all his motions.

The notes of the Wren are very lively and garrulous, and if not uttered more frequently during the heat of the day, are, on account of the general silence of birds, more noticeable at that hour. There is a concert at noonday, as well as in the morning and evening, among the birds; and of the former the Wren is one of the principal musicians. After the hot rays of the sun have silenced the early performers, the Song-Sparrow and the Red-Thrush continue to sing at intervals during the greater part of the day. The Wren is likewise heard at all hours; but when the languishing heat of noon has arrived, the few birds that continue to sing are more than usually vocal, and seem to form a select company. The birds which are thus associated with the Wren are the Bobolink, the Preacher, the Linnet, and the Catbird, if he be anywhere near. If we were at this hour in the woods we should hear the loud, shrill voice of the Oven-Bird and some of the warbling sylvians.

Of all these noonday singers, the Wren is the most remarkable. His song is singularly varied and animated. He has great compass and execution, but wants variety in his tones. He begins very sharp and shrill, like a grasshopper, slides down to a series of guttural notes, then ascends like the rolling of a drum in rapidity of utterance to another series of high notes. Almost without a pause he recommences his querulous insect-chirp, and proceeds through the same trilling and demi-semi-quavering as before. He is not particular about the part of his song which he makes his closing note. He will leave off in the middle of a strain, when he seems in the height of ecstasy, to pick up a spider or a fly. As the Wren produces two broods in a season, his notes are prolonged to a late period in the summer, and may be heard sometimes in the third week in August.

### THE WINTER-WREN.

We do not often meet with this bird near Boston in summer. He is then a resident of the northern parts of Maine and New Hampshire, and of the Green Mountain range. In the autumn he migrates from the north and may be occasionally seen in company with our other winter birds. In our own latitude, if the cold season drives him farther south, we meet him again early in the spring, making his journey to his northern home. While he remains with us we see him near the shelving banks of rivers, creeping about old stumps of trees, which, half decayed, furnish a frugal share of his dormant insect-food. He is so little afraid of man that he will often leave his native resorts, and may be seen, like our common House-Wren, examining the wood-pile, creeping into the holes of old stone-walls and about the foundations of out-houses. Not having seen this bird except in winter, I am unacquainted with his song. Samuels describes it as very melodious and delightful.

### THE MARSH WREN.

I was once crossing by turnpike an extensive meadow which was overgrown with reeds and rushes, when my curiosity was excited by hearing, in a thicket on the banks of a streamlet, a sound that would hardly admit of being described. I could not tell whether it came from an asthmatic bird or an aggravated frog. The sound was unlike anything I had ever heard. I should have supposed, however, if there were Mocking-Birds in our woods, that one of them had concealed himself in the thicket and was attempting to imitate the braying of an ass. I sat down upon the railing of a rustic bridge that crossed the stream, and watched for a sight of the imp that must be concealed there. In less than a minute there emerged from it a Marsh-Wren, whisking and flitting about with gestures as peculiar though not as awkward as his burlesque song.

If I believed, as some writers affirm, that birds learn their song from their parents, who carry them along from one step to another as if they had a musical gamut before them, I might have conjectured that this bird had been taught by a frog, and that, despising his teacher, he strove not to learn his reptile notes but to burlesque them. As I was walking homeward, I could not but reflect that Nature, who is sometimes personified as an old dame, must have indulged her mirthfulness when she created a bird with the voice of a reptile.

Dr. Brewer describes the nest of the Marsh-Wren as nearly spherical, composed externally of coarse sedges firmly interwoven, cemented with mud and clay, and impervious to the weather. An orifice is left on one side for entrance, having on the upper side a projecting edge to protect it from rain. The inside is lined with soft grass, feathers, and the cottony product of various plants. It is commonly placed on a low bush a few feet from the ground.

This species, like all the Wrens, has great activity and industry, consumes immense quantities of small insects, is very petulant in its manners, and manifests a superior degree of intelligence and courage.

# THE HAUNTS OF FLOWERS.

THERE is not a more interesting subject connected with botany than that of the haunts of flowers. We may by chance discover a rare and beautiful plant in a situation that would be the last to invite our attention. The apparent unfitness of the place for aught but common weeds may have preserved it from observation. I have sometimes encountered by the roadside a species for which I had long vainly traversed the woods. On the borders of some of the less frequented roads in the country, the soil and the plants still remain in their primitive condition. In such grounds we may find materials for study for several weeks, without leaving the waysides. Indeed, all those old roads which are not thoroughfares — by-ways not travelled enough to destroy the grass between the ruts of wheels and the middle path made by the feet of horses — are very propitious to the growth of wild plants. The shrubbery on these old roadsides, when it has not been disturbed for a number of years, is far more beautiful than the finest imaginable hedgerow. Here are several viburnums, two or three species of cornel, the bayberry, the sweet fern, the azalea, the rhodora, the small kalmia, and a crowd of whortleberry-bushes, beside the wild rose and eglantine. The narrow footpath through this wayside shrubbery has a magic about it that makes it delightful to pass through it. Under the shelter of this tangle-wood Nature calls out the anemone, blue, white, and pedate violets, and in damp places the erythronium, the Solomon's seal, and the bellwort. When I see these native ornaments destroyed for the improvement of the road, I feel like one whose paternal estate has been cleared and graded and measured out into auction-lots.

There are indications by which we may always identify the haunts of certain species, if they have not been eradicated. We know that fallow grounds are inhabited by weeds, and that mean soils contain plants that seem by their thrift to require a barren situation; but they are like poor people, who live in mean huts because the better houses are occupied by their superiors. These plants would grow more luxuriantly in a good soil, if they were not crowded out by those of more vigorous habit. Every one is familiar with a species of rush called wire grass, which is abundant in footpaths through wet meadows. It is so tough that the feet of men and animals, while they crush and destroy all other plants that come up there, leave this uninjured. This remarkable habit has caused the belief that it thrives better from being trampled under foot. The truth is, it will bear more hard usage than other species, and is made conspicuous by being left alone after its companions have been trodden to death. The same may be observed of a species of *Polygonum*, — the common "knot-grass" of our back yards. A certain amount of trampling is favorable to its growth by crushing out all its competitors.

Most of our naturalized plants inhabit those places which have been once reduced to tillage and afterwards restored to nature. Such are the sites of old gardens and orchards, and the forsaken enclosures of some old dwelling-house. The white Bethlehem-star is a tenant of these deserted grounds, growing meekly under the protection of some moss-covered stone-wall or dilapidated shed, fraternizing with the celandine, the sweet chervil, and here and there a solitary narcissus. The euphorbia and houseleek prosper in similar places, growing freshly upon ledges and heaps of stones, which have been carted by the farmer into abrupt hollows, mixed with the soil and weeds of the garden. In shady corners we find the coltsfoot, the gill, — a very pretty labiate, — and some of the foreign mints. Spikenard and tansy delight in more open places, along with certain other medicinal herbs introduced by ancient simplers. These plants are seldom found in woods or primitive pastures.

Wild plants of rare beauty abound in a recent clearing, especially in a tract from which a growth of hard wood has been felled, if afterwards the soil has remained undisturbed. In the deep woods the darkness will not permit any sort of undergrowth except a few plants of peculiar habit and constitution. But after the removal of the wood, all kinds of indigenous plants, whose seeds have been wafted there by the winds or carried there by the birds, will revel in the clearing, until they are choked by a new growth of trees and shrubs. Strawberries and several species of brambles spring up there as if by magic, and cover the stumps of the trees with their vines and their racemes of black and scarlet fruit; and hundreds of beautiful flowering plants astonish us by their presence, as if they were a new creation. We must look to these clearings, and to those tracts in which the trees have been destroyed by fire, more than to any others, for the exact method of nature. Among the first plants that would appear after the burning, beside the liliaceous tribe, whose bulbs lie too deep in the soil to be destroyed, are those with downy seeds, which are immediately sowed there by the winds. One very conspicuous and beautiful plant, the spiked willow herb, is so abundant in any tract that has been burned, the next year after the conflagration, that in the West and in the British Provinces it has gained the name of fireweed.

## THE HAUNTS OF FLOWERS.

But the paradise of the young botanist is a glade, or open space in a wood, usually a level between two rocky eminences, or a little alluvial meadow pervaded by a small stream, open to the sun, and protected at the same time from the winds by surrounding hills and woods. It is surprising how soon the flowery tenants of one of these glades will vanish after the removal of this bulwark of trees. But with this protection, the loveliest flowers will cluster there, like the singing-birds around a cottage and its enclosures in the wilderness. Here they find a genial soil and a natural conservatory, and abide there until some accident destroys them. Nature selects these places for her favorite garden-plots. In the centre she rears her tender herbs and flowers, and her shrubbery in the borders, while the trees form a screen around the whole. I have often seen one of these glades crimsoned all over with flowers of the cymbidium and arethusa, with wild roses in their borders, vying in splendor with a sumptuous parterre.

While strolling through a wood in one of those rustic avenues which have been made by the farmer or the woodman, we shall soon discover that this path is likewise a favorite resort for many species of wild-flowers. Except the glade, there are but few places so bountifully stored by nature with a starry profusion of bloom. The cranesbill, the wood anemone, the cinquefoil, the yellow Bethlehem-star, the houstonia, to say nothing of crowds of violets, adorn the verdant sward of these woodpaths; and still beyond them, cherished by the sunshine that is admitted into this opening, ginsengs, bellworts, the white starlike trientalis, the trillium, and medeola thrive more prosperously than in situations entirely wild and primitive. It is pleasant to note how kindly Nature receives these little disturbances which are made by the woodman, and how many beautiful things will assemble there, to be fostered by those conditions which accident, combined with the rude operations of agriculture, alone can produce.

Leaving this avenue, we ascend the sloping ground, and, passing through a tangled bed of lycopodiums, often meeting with the remnants of a foot-path that is soon obliterated in a mass of vegetation; then wandering pathless over ground made smooth by a brown matting of pine leaves, beautifully pencilled over with the small creeping vines and checkered foliage of the mitchella and its scarlet berries, we come at last to a little rocky dell full of the greenery of mosses and ferns, and find ourselves in the home of the columbines. Such a brilliant assemblage reminds you of an aviary full of linnets and goldfinches. The botanist does not consider the columbine a rare prize. It is a well-known plant, thriving both in the wood and outside of it; but it is gregarious, and selects for its habitation a sunny place in the woods, upon a bed of rock covered with a thin crust of soil. The plants take root on every rocky projection and in every crevice, hanging like jewels from a green tapestry of velvet moss.

As we leave this magic recess of flowers and pursue our course under the pines, trampling noiselessly over the brown, elastic sward, we soon discover the purple, inflated blossoms of the pink lady's-slipper. These flowers are always considerably scattered, and never grace the open field. Often in their company we observe the sweet pyrola, bearing a long spike of white flowers that have the odor of cinnamon. Less frequently we find in this scattered assemblage some rare species of wood orchis and the singular coral plant. If we now trace the course of any little streamlet to a wooded glen full of pale green bog-moss, covering the ground with a deep mass of spongy vegetation, there we may happily discover the rare and beautiful white orchis, the nun of the woods, with flowers resembling the pale face of a lady wearing a white cap. This plant is found only in certain cloistered retreats, under the shade of trees. It is a true vestal, and will not tarnish its purity by any connection with the soil. It is cradled like an infant in the soft, green bog-moss, and derives its sustenance from the pure air and dews of heaven. Like the orchids of warm climates, it is half parasitic, and requires certain conditions for its growth which are rarely combined.

Flowers are usually abundant in pleasant situations. They avoid cold and bleak exposures, the dark shade of very dense woods, and wet places seldom visited by sunshine. Like birds, they love protection, and we are sure to meet with many species wherever the songsters of the forest are numerous. Birds and flowers require the same fostering warmth, the same sunshine, and the same fertility of soil to supply them with their food. When we are traversing a deep forest, the silence of the situation is one of the most notable circumstances of our journey; but if we suddenly encounter a great variety of flowers, our ears will at the same time be greeted by the notes of some little thrush or sylvia. If I hear the veery, a bird that loves to mingle his liquid notes with the sound of some tuneful runlet, I know that I am approaching the shady haunts of the trillium and the wood thalictrum. If I hear the snipe feebly imitating the lark, as he soars at twilight, and warbles his chirruping song far above my head, I know that when he descends in his spiral course he will alight upon grounds occupied by the Canadian rhodora, the andromeda, and the wild strawberry plant. But if the song of the robin is heard in the forest, I feel sure that a cottage is near, with its orchard and cornfields, or else that I am close to the end of the wood and am about to emerge into the open plain.

A moor is seldom adorned with plants that would prosper in the uplands; but if it be encompassed by a circle of wooded hills, a gay assemblage of flowers will congregate in its borders, where hill and plain are imperceptibly blended. We may always find a path made by cattle all along the border. If we thread the course of this path, we pass through bushes of moderate height, consisting of whortleberries, clethra, and swamp honeysuckles, and now and then enter a drier path, through beds of sweet-fern, and occasional open spaces full of pedate violets. The docile animals, — picturesque artists who constructed this path, — while gazing upon the cloverpatches, will turn their large eyes placidly upon us, still heeding their diligent occupation. We keep close to the edge of the moor, not disregarding many common and homely plants that lie in our way, till we discover the object of our search, the sarracenia, or sidesaddle plant, with its dark purple flowers, nodding like Epicureans over their circle of leafy cups half filled with dew. This is a genuine "pitcher plant," and is the only one of the family that is not tropical. The water avens — conspicuous for its drooping chocolate-colored flowers — and the golden senecio congregate in the same meadow, bending their plumes above the tall rushes and autumnal asters not yet in flower.

Very early in the season, if we are near an oak wood, standing on a slope with a southern exposure, we enter it, and if fortune favors us, the liverwort will meet our sight, pushing up the dry oak leaves that formed its winter covering, and displaying its pale bluish and purple flowers, deepening their hues as they expand. When they are fully opened, there are but few sights so pleasant as these circular clusters of flowers, on a ground of dry brown foliage, enlivened with hardly a tuft of verdure, except the trilobate leaves of this interesting plant. As oaks usually stand on a fertile soil, there is a greater variety of species among their undergrowth than in almost any other wood. A grove of oaks, after it has been thinned by the woodman so as to open the grounds to the sun, becomes when left to nature a rare repository of herbaceous plants. Yet there are certain curious species which are found almost exclusively in pine woods. Such is the genus Monotropa, including two species, the pine-sap and the bird's-nest, — plants without leaves or hues, with stems resembling potato-sprouts grown in a dark cellar. Outside of pine woods, however, on their southern boundary, we may always look for the earliest spring flowers, because no other wood affords them so warm a protection.

In our imaginary tour we have visited only the most common scenes of nature; we have traced to their habitats very few rare plants, and have yet hardly noticed the flowers of autumn, — those luxuriant growers, many of them half shrubby and branching like trees. Some of these have no select haunts. The asters and golden-rods, the most conspicuous of the hosts of autumn, are found in almost every soil and situation; but they congregate chiefly on the borders of woods and fields, and seem to take special delight in arraying themselves by the sides of roads recently laid out through a wet meadow. The autumnal plants generally prosper only in the lowlands which have not suffered from the summer droughts. When botanizing at the close of the season, we must avoid dry sandy places, and follow the windings of narrow streams that glide through peat-meadows, and traverse the sides of ditches, examining the convex embankments of soil which have been thrown up by the spade of the ditcher. On level moors we meet with occasional rows of willows affectionately guarding the waters of these artificial pools, where they were planted as sentinels by the rustic laborer. The gentians, which have always been admired, as much for the delicacy and beauty of their flowers as for their hardy endurance of autumnal frosts, are often strewn in these places, glowing like sapphires on the faded greensward of the closing season of vegetation.

The great numbers of wild plants which are often assembled in a single meadow seem to a poetical mind something more than a result of the mere accidents of nature. There is not a greater variety or diversity in the thoughts that enter and pass through the mind than of species among these herbs. Each has distinct features, and some attractive form or color, or other remarkable property peculiar to itself. How many different species bend under our footsteps while we are crossing an ordinary field! How many thousands are constantly distilling odors into the atmosphere, which is oxygenated by their foliage and purified and renovated by their vital and chemical action! There is not a single plant, however obscure, minute, or unattractive, that is not an important agent of Nature in her vast and mysterious economy.

There would be no end to our adventures, if we were resolved to continue them until our observations were exhausted. Hence the never-failing resources of the botanist who is within an hour's walk of the forest. The sports of hunting and fishing offer their temptations to a greater number of young persons; but they do not afford continued pleasure to their votaries, like botanizing. The hunter watches his dog and the angler his line; but the plant-hunter examines everything that bears a leaf or a flower. His pursuit leads him into all the green recesses of nature, — into sunny dells and shady arbors, over pebbly hills and plashy hollows, through mossy dingles and wandering footpaths, into secret alcoves where the Hamadryads drape the rocks with ferns, and Naiads collect the dews of morning and pour them into their oozy fountains for the perfection of their verdure.

A ride over the roads of the same region is not so pleasant as these intricate journeys of the botanist. He fraternizes with all the inhabitants of the wood, and with the laborers of the farms which he crosses, not heeding the cautions to trespassers. He meets the rustic swain at his plough, and listens to his quaint discourse and his platitudes about nature and mankind. He follows the devious paths of the ruffed grouse, and destroys the snares which are set for its destruction. He listens to its muffled drum while he cools his heated brow under a canopy of maples overarched with woodbine, and picks the scarlet berries that cluster on the green knolls at his feet. He lives in harmony with all created things, and hears the voices of the woods and music of the streams. The trees spread their shade over him; every element loads him with its favors. Morning hails him with her earliest salutation, and introduces him to her fairest hours and sweetest gales. Noon tempts him into her silent woodland sanctuaries, and makes the hermit thrush his solitary minstrel. Evening calls him out from his retreat, to pursue another varied journey among the fairy realms of vegetation, and ere she parts with him curtains the heavens with splendor and prompts her choir of sylvan warblers to salute him with their vespers.

# WATER SCENERY.

THERE is no single thing in nature that adds more beauty to landscape than water. It is emblematical of purity and tranquillity; it is suggestive of multitudes of pleasant rural images, and, beside these moral expressions, it possesses a great deal of intrinsic beauty. The mirrored surface of a lake or a stream, reflecting the hues and forms of the clouds in the heavens, and of the trees and shrubbery on its banks, is pleasing to the eye, independently of any suggestion that may occur to a fanciful mind. The eye requires to be practised, or rather the mind must be educated in a certain manner, before it can enjoy and appreciate moral beauty. But the beauty of a smooth surface of water, of waves trembling in the moonlight, of a spouting fountain, or a sparkling rill, is obvious and attractive even to a child. In water have color and form and motion intimately combined their charms, assuming the loveliest tints in the dews of heaven and the spray of the ocean, and every imaginable form of beauty in the lake and its sinuosities, and the river in its various windings through vale and mountain.

Water is not only beautiful in itself, but it is one of the chief sources of pleasing variety in the expression of landscape, whether we view it as spread out on the silver bosom of a lake, the serpentine course of a river, or by its outlines forming those endless changes that delight the voyager by the sea-shore. Every one must have observed, when riding through an unattractive country, how it seems overspread with a sudden charm when we come in sight of a lake or stream. What was before monotonous is now agreeably varied; what before was spiritless is now animated and cheering. A similar effect is produced by the sight of a little cottage in a desert or uninhabited region, or in the midst of an uncultivated plain. The eye wanders about unsatisfied, until it sees this human dwelling, when it rests contented, because it has found something to fix the attention and to awaken a sympathetic interest. We are not always aware how greatly the beauty of landscape is founded on our habitual associations. At the sight of water we think at once of the numerous delights, bounties, and luxuries that flow from its beneficent streams; and perhaps nothing in a prospect so instantaneously awakens the idea of plenty and of the beneficence of nature. Water is, therefore, the very picture of benevolence, without which the face of the country would seem cold, ungenerous, and barren.

A feeling of seclusion is one of the agreeable emotions connected with a ramble in the woods; and some delightful spots derive their principal attractions from their evident adaptedness to this security from observation. When we are walking, either alone or in company, we do not like to be met by others or to be observed by them. A little sequestered spot, that seems to offer all this desirable shelter from the eyes of the world, is always singularly attractive. But those are the most eligible retreats in which one might be secluded and at the same time accommodated with a pleasant and extensive prospect. To be able to look out upon the world from a little nook, while unobserved and not liable to be interrupted by others, affords one an experience of the same emotion with which we contemplate the raging of a storm from a place of comfort and security.

Water is in a high degree favorable to the attainment

of these pleasant advantages. Let two parties be placed at opposite points, with a small lake intervening, and though full in sight of each other, they still feel secluded. The pleasantness of their retreat, under these circumstances, is enlivened by the sight of the opposite party, and they may be amused by observing the motions of the other, and at the same time feel secure from intrusion. But if there were only a meadow of equal width to separate them, the secluded character of the situation would be lost; as the parties are not only in sight, but are liable to be interrupted by a visit from the opposite one. A lake may in this way be the occasion of many of those delightful retreats, attended with advantages of prospect, which no other combination of scenes could so well afford. The beauty of many of these situations depends greatly on their apparent adaptedness to this kind of recreation and seclusion.

A river, especially of moderate width, is in many respects more beautiful than a lake; and, more than any other collection of water, suggests the idea of infinity and of continued progression. I never look upon a clear stream of narrow dimensions, without thinking of the thousand beautiful scenes it must visit, in its blue course through the hills and plains. What a life of perpetual delight must be led by the gentle river goddess, as she is wafted up and down the stream in her shallop of reeds! Now coursing along under banks sprinkled over with honeysuckles, while their fragrance follows the current of the stream, to entice the bees and other insects to their fragrant flower-cups; then passing through a pleasant forest, where she is regaled by the terebinthine odor of pines mingled with that of flowering lindens, whose branches resound all day with the hum of insects and the warbling of birds. Every green bank offers to her hand a profusion of wild strawberries, and every rocky declivity hangs its brambles over the stream, and tempts her with delicate clusters of raspberries, and other delicious fruits. How, if she takes pleasure in the happiness of human beings, must she be charmed by witnessing the plenty which is everywhere diffused by the crystal waters of her own stream; the countless farms rendered fertile and productive through its agency; the numerous mill-seats that derive their power from its falls and rapids, and gather the industrious inhabitants in smiling hamlets upon its banks! A river, when pursuing its winding course along the plain, alternately appearing and disappearing among the hills and woods, suggests the idea of a pleasant journey, and is peculiarly emblematical of human progress. It always seems to me that it must conduct one to some happier region, and that if I traced it to its source, I should be led into the very temple of the Naiads!

With the different forms of water are associated nearly all the pleasant images of rural life. To one who is tired of his busy employments in the city, a rural retreat is like a cool breeze to the traveller in a sultry desert. A little arbor, that overlooks a river, a lake, or an arm of the sea, derives its charms almost wholly from the water, which is at the same time the symbol of peace and plenty, and the mirror of heaven. A hermitage by the side of a stream affords a secret retreat, still more delightful from its fancied association with pious seclusion from the world. Every flower that looks up to us from the green, mossy turf; every bird that warbles in the neighboring copse; and every insect that hums in the herbage at our feet, has a soothing influence, that for a season dispels every care and every feverish excitement. Then do we feel that nature only has power to administer that solace which is balm to the soul when one is vexed with care and weary of men.

One of the sentiments often awakened by a water prospect is that of sublimity. But this can only arise from an extensive view of the ocean or of a cataract. Ordinarily, therefore, except by the sea-shore, we seldom behold a sufficient expanse of water to affect us with this sentiment. Its influence is greater when a wide sea-view comes suddenly upon the eye, after one has passed through a succession of beautiful, quiet, and rather confined scenes. Small lakes and rivers greatly enhance the beauty of a pastoral landscape, because they afford evidence of good pasturage and a plentiful supply of water to the flocks and grazing herds. Painters, taking advantage of this expression, often represent, in one of their side views, the cattle standing up to their knees in a little pond of water, while the green rushes and undefaced shrubbery growing about them make manifest its clearness and purity. Ocean scenery is not favorable to pastoral expression; but it enhances the beauty of sunrise, and adds grandeur to the sublimity of a tempest.

Many writers have eulogized an ocean prospect as beheld from a point where we can see no land. The views presented by the ocean from different points on the shore, which is broken and intersected by frequent inlets of water, we can never cease to admire; but I have little sympathy with these lovers of boundless space. The eye soon tires of gazing upon a scene that awakens no emotion but that of infinity, and presents no point as a resting-place for the imagination. To the sublimity of an ocean voyage, with its mountainous waves and its interminable azure, I prefer a boat excursion on a narrow stream, where the trees on the opposite banks frequently interlace their branches over the middle of the current, and the plashing of the oar often startles the little twittering sandpiper that is feeding upon the edge of the stream. The sight of a small lake surrounded by woods, and dotted all round its borders with full-blown waterlilies, over whose broadspread leaves the little plover glides, without impressing a ripple on the glossy brink, gives me more pleasure than I could derive from any view of the ocean, bounded only by the horizon.

Water needs the accompaniment of field and wood to form a picture that is agreeable to the eye. Without such adjuncts, it is like the sky when it has no clouds, and is void of all pleasing suggestions. The pleasure of angling on the banks of a river or a lake is greatly magnified by the agreeable combination of wood and water scenery that surrounds us. The beauty of an island is like that of a lake; and it is hard to say which of the two affects the spectator with the most delight, though I am inclined to believe that the majority would decide in favor of the island. The island, especially if there be a little cottage upon it, is suggestive of a multitude of pleasing fancies connected with rural life and retirement. In this case, we think not so much of the difficulty of gaining access to town, or even of coming on shore, as of the peaceful seclusion it seems calculated to afford. The lake suggests no such ideas; it is chiefly attractive by its own beautiful sheen of crystal water, by its association with boat excursions on serene afternoons or moonlight evenings, and with rural pleasures connected with the scenes on its shore.

# APRIL

DEAR to the poet and to the lover of nature is the month of April, when she first timidly plants her footsteps upon the dank meadow and the mossy hillside, clothing the dark brown sods with tufts of greenery, waking the early birds, and cherishing the tender field-flowers. Her hands are ever busy, hanging purple fringes upon the elm and golden tassels upon the willow bough, and weaving for the maple a vesture of crimson. She brings life to the frozen streams, verdure to the seared meadows, and music to the woods, which have heard nothing for months save the solemn moaning of their own boughs and the echoes of the woodman's axe from an adjoining fell. We welcome April as the comforter of our weariness after long confinement, as the bearer of pleasures which her bounty only can offer, as a sweet maiden entering the door of our prison with hands full of budding flowers and breath scented with violets.

A gladness and hopefulness attend us on the return of spring which are unfelt at other seasons, and produce a sensation like that of the renewal of youth. We are certainly more hopeful at this time than in the autumn, and we look back upon the lapse of the three winter months with a less painful sense of the loss of so much of our allotted period of life than upon the lapse of the three summer months. Though the flight of any season carries us equally onward in our mortal progress, we cannot avoid the feeling that the lapse of winter is our gain as that of summer was our loss. And surely, of these two reflections, the one that deceives is better than the one that utters the truth; and though we are several months older than we were in the autumn, we may thank Heaven for the delusion that makes us feel younger.

Spring, the true season of hopefulness and action, is unfavorable to thought. So many delightful objects are constantly inviting us to pleasure, that we are tempted to neglect our serious pursuits, and we feel too much exhilaration for confinement or study. It is not while surrounded by pleasures of any kind that we are most capable of reflecting upon them or describing their influence; for the act of thinking upon them requires a suspension of our enjoyments. Hence, in winter we can most easily discourse upon the charms of spring and summer, when the task becomes a pleasant occupation, by reviving the scenes of past delights blended with a foretaste of joys that are to come. But when the rising flowers, the perfumed breezes, and the music of the animated tenants of the streams, woods, and orchards, are all inviting us to come forth and partake of the pleasures they proffer, it is wearisome to sit down apart from all these delights to the comparatively dull task of describing them.

As childhood is not always happy, and as youth is liable to the sorrows and afflictions of later life, the spring is not always cheerful, and the vernal skies are sometimes blackened with wintry tempests, and the earth bound in ice and frost. Even in April the little flowers that are just peeping out from their winter coverts are often greeted by snow, and spring's "ethereal mildness" is exchanged for harsh winds and cloudy skies. In vain do the crocus, the snowdrop, and the yellow narcissus appear in the gardens, or the blue violet and the saxifrage spangle the southern slopes of the hills,— the north-wind is not tempered by their beauty nor beguiled by the songs of the early birds.

April — the morning of the year, as March was its twilight, — that uncertain time when the clouds seem like exiled wanderers over the blue field of light, hurrying in disorganized cohorts to some place of rest or dissolution — daily flatters us with hopes which she seems reluctant to fulfil. But every invisible agent of nature is silently weaving a drapery of verdure to spread around the footsteps of the more lovely month that is soon to arrive. We see the beginnings of this work of resurrection in thousands of small tufted rings of herbage scattered over the fields, and daily multiplying, until every knoll is crowned with blue, white, and crimson flowers that will join to gladden the heyday of spring.

When at length the south-wind calls together his vernal messengers, and leads them forth in the sunshine to their work of gladness, the frosty conqueror resigns his sceptre, and beauty springs up in the place of desolation. The bee rebuilds his honeyed masonry, the swelling buds redden in the maples, and every spray of the forest and orchard is brightened with a peculiar gloss that gives character to the vernal tinting of the woods. The ices that have bound the earth for half the year are dissolved; the mountain snows are spread out in fertilizing lakes upon the plains, and the redwing pipes his garrulous notes over the abiding-place of the trillium and the meadow cowslip. The lowlands, so magnificent in autumn, when glowing with a profusion of asters and golden-rods, are now whitened with this sheet of glistening waters, put into constant agitation by multitudes of frogs tumbling about in the shallows while engaged in their croaking frolics.

April is the month of brilliant skies constantly shadowed by dark, rapidly moving clouds, of brown meadows and plashy foot-paths. The barren hills are velveted with moss of a perfect greenness, delicately shaded with a profusion of glossy purple stems, like so many hairs, terminated with the peculiar flower of the plant; and long stripes of verdure mark the progress of the new-born rivulets, as they pursue their irregular course down the hillside into the valleys. But the damp grounds, frequently almost impassable from standing water, are interspersed with little dry knolls covered with mosses and lycopodiums, where the early flowers of spring delight to nestle, embosomed in their soft verdure. Upon these evergreen mounds the fringed polygala spreads a beautiful hue of crimson; and while gathering its flowers, we discover, here and there, a delicate wood-anemone, with its mild eyes not yet open to the light of day. But so few flowers are abroad that the bee when it comes forth in quest of honey must feel like one who is lost and wandering in space. It can revel only in gardens where the sweet-scented flowers of another clime spread abroad a perfume that is but a false signal of the weather of its adopted climate.

The odors that perfume the air in the latter part of this month are chiefly exhaled from the unfolding buds of the flowering trees and shrubs, and from pine woods. The balm of Gilead and other poplars, while the scales are dropping from their hibernacles, to loose the young leaves and flowers from their confinement, afford the most grateful of odors, and are a part of the peculiar incense of spring. But there are exhalations from the soil in April, when the ploughman is turning his furrows, that afford an agreeable sensation of freshness, almost like fragrance, resembling the scent of the cool breezes, which, wafted over beds of dulses and sea-weeds, when the tide is low, often rise up suddenly in the heat of summer.

As April advances, the familiar bluebirds are busy among the hollows of old trees, where they rear their young secure from depredation. Multitudes of them, seen usually in pairs and seldom in flocks, are distributed over the orchards, responding to one another in their few plaintive, but cheerful, notes; and their fine azure plumage is beautifully conspicuous as they flit among the branches of the trees. The voice of the robin resounds in all familiar places, and the song of the linnet is heard in the groves which have lately echoed but with the screaming of the jay and the cawing of the raven. Young lambs, but lately ushered into life, may be seen with various antic motions, trying the use of their limbs, that seem to run wild with them before they have hardly ascertained their powers; and parties of little children, some with baskets, employed in gathering salads, others engaged in picking the scarlet fruit of the checkerberry, will often pause from their occupations with delight to watch the frolics of these happy creatures.

The small beetles that whirl about on the surface of still waters have commenced their gambols anew, and fishes are again seen darting about in the streams. A few butterflies, companions of the crocus and the violet, are flitting in irregular courses over the plains; the spider is hanging by his invisible thread from the twigs of the orchard trees, and insects are swarming in sunny places. The leaves of the last autumn, disinterred from the snow, are once more rustling to the winds and to the leaping motions of the squirrel. Small tortoises are basking in the sunshine upon the logs that extend into the pool; and as we draw near we see their glistening armor, as with awkward haste they plunge into the water. The ices which had accumulated around the sea-shore have disappeared, and the little fishes that congregate near the edges of the salt-water creeks make a tremulous motion of the water, as upon our sudden approach they dart away from the shallows into the deeper sea.

The sun has sunk below the horizon. The wind is still,

and the countless lakes that cover the meadows reflect from their mirrored surfaces an image of every cloud that floats above them. The bright-eyed evening star now shines alone. The lowing of cattle is heard only at intervals from the farmyards, and the occasional sound of distant bells is borne softly in the hush of day's decline. The birds are silent in the woods, save now and then a solitary one, greeted perhaps by a lingering sunbeam reflected from a radiant cloud, will sing a few twittering notes of gladness. But nature is not silent. The notes of myriads of little piping musicians rise in a delightfully swelling chorus, from every lake and stream, now loudening with an increased multitude of voices, then dying away into a momentary silence. These sounds are the charm of an April evening; and in my early days I listened to them with more pleasure than to the sweetest strains of music, as prophetic of the reviving beauties of nature. And now, when the first few piping notes fall upon my ear, my mind is greeted by a vision of dearly remembered joys that crowd vividly upon the memory. These tender recollections, blended with the hopes and anticipations of spring, serve with peculiar force to tranquillize the mind and render it cheerful and satisfied with the world.

# THE PLUMAGE OF BIRDS.

THE colors and forms of the plumage of birds are generally regarded as mere accidents, unattended with any advantages in their economy. I cannot believe, however, that they are not in some way, which we cannot fully understand, indispensable to their existence as a species. Let me then endeavor to discover, if possible, the design of Nature in spreading such a variety of tints upon the plumage of birds, and to learn the advantages they derive from these native ornaments. Do they affect the vision of birds with the sensation of beauty, and serve to attract together individuals of the same species? Or are they designed also to protect them from the keen sight of their enemies, while flitting among the blossoms of the trees? It is probable that each of these purposes is subserved by this provision of Nature. She has clothed individuals of the same species and the same sex with uniformity, that they may readily identify their own kindred, and has given them an innate susceptibility to derive pleasure from those colors that predominate in the plumage of their own species. She has likewise distinguished the small birds that live on trees by beautiful colors, while those in general that run upon the ground are marked by neutral tints, that the former may be less easily observed among the blossoms of the trees, and that the latter may be less conspicuous while sitting or running upon the ground.

It is well known that the males of many species are more beautifully and brilliantly decorated than the females, and that the singing-birds in general have less beauty of color than the unmusical species. As an explanation of this fact we must consider that the singing-birds are more humble in their habitats than others. The brightly colored birds chiefly frequent the forests and lofty trees. Such are the woodpecker, the troupial, and many species of tropical birds. The northern temperate latitudes are the region of the grasses, which afford sustenance to a large proportion of the singing-birds — the finches and buntings — of that part of the world. Some of the finches are high-colored, but these usually build in trees, like the purple finch and the goldfinch. But the sparrows and the larks, that build in a bush or on the ground, are plainly dressed. The thrushes, which are equally plain in their dress, build in low bushes, and take their food chiefly from the ground. Indeed, it might be practicable to distinguish among a variety of strange birds the species that live and nestle in trees by their brighter plumage.

In our own latitude the species that frequent the shrubbery are of a brown or olive-brown of different shades. They are dressed in colors that blend with the general tints of the ground and herbage while they are seeking their food or sitting upon their nests. Birds, however, do not differ much in the colors of the hidden parts of their plumage. Beneath they are almost universally of grayish or whitish tints, so that, while sitting on a branch, the reptiles lurking for them may not, when looking upward, distinguish them from the hues of the clouds and the sky and the grayish undersurface of the leaves of trees. Water-birds are generally gray all over, except a tinge of blue in their plumage above. Ducks, however, are many of them variegated with green and other colors that harmonize with the weeds and plants of the shore upon which they feed.

Nature works on the same plan in guarding insects

and reptiles from the sight of their foes. Thus, the toad is colored like the soil of the garden, while the colors of the common frog that lives among the green rushes and aquatic mosses are green, and the tree-frog is of a mottled gray, like the outer bark of old trees. Grasshoppers are generally greenish; but there is a species found among the gray lichens on our rocky hills which is the color of the surface of these rocks.

Among the singing-birds of this country which are remarkable for their brilliant colors are the golden oriole, the scarlet tanager, and the American goldfinch. All these species build their nests in trees, and seldom run on the ground. The goldfinch feeds upon the seeds of compound flowers, which are mostly yellow. His plumage of gold and olive allows him to escape the sight of an enemy while picking seeds from the disk of a sunflower or from a cluster of goldenrods.

But why are the females plainly dressed and the males alone adorned with brilliant colors? It may be answered, that, as the female performs the duties of incubation, if she were brightly colored like the male, she would be more readily descried by a bird of prey while sitting on her nest. The male, on the contrary, while hunting among the blossoms and foliage of trees for his insect food, is not so readily distinguished from the flowers, for in temperate latitudes the breeding season is the time when the trees are in blossom. After the young are reared and the flowers have faded, several species dispense with their brilliant colors and assume the plain hues of the female.

We must consider, however, that the beautiful colors of the plumage of the male birds serve to render them more conspicuous objects of attraction to the females. Hence, in the early part of the year, just before the time of courtship arrives, Nature has provided that the plumage of various kinds of birds should suffer a metamorphosis. Thus the bobolink exchanges his winter garment of yellowish-brown for one of brilliant straw-color and black; and the red-winged blackbird casts off his tawny suit for one of glossy jet, with epaulettes of scarlet. What are the useful ends subserved by this mysterious provision of Nature? She clothes them with beauty and endows them with song at a period when their success as lovers depends greatly on the multitude and power of their attractions. Among the beautiful species their success is in proportion to the splendor of their plumage; and among the warblers, to the charms of their voice. Beauty and song are the means Nature has furnished them, whereby they may render themselves, I will not say agreeable, but attractive. I do not suppose a beautiful male bird is preferred to a plain one of the same species; but his beauty causes him to be sooner discovered by an unmated female.

It is easy to explain, therefore, on the principle of compensation, why handsome birds in general are endowed with inferior musical powers. They are able to accomplish by their beauty of plumage what the plainer species do by their songs. It may be observed that the handsome birds, when engaged in courtship, place themselves in attitudes which are calculated to display the full beauty of their plumage; while the songsters under the same circumstances pour forth an unusual strain of melody. The hues of the brightly colored male birds may be a means of assisting their young in identifying them after they have left their nest. They hear, for example, the loud call-note of the golden robin, and immediately they recognize him by his colors, when, if plainly dressed, they might not discover him. As soon as they behold him they commence their chirping and are greeted by the old bird.

There is one numerous tribe of birds that run upon

the ground, whose males, except those of a few species, are very brilliantly decorated. This is the gallinaceous family, which are an exception to my remark that the handsome birds inhabit trees. But it is only the larger species or genera of this family, such as the pheasant, the turkey, the peacock, the curassow, and the common fowl, whose males are thus gorgeously arrayed. Their colors are evidently intended for their protection in a peculiar way. All the males of these species are endowed with a propensity to ruffle and expand their feathers whenever they are threatened with attack. The boldest animal would be frightened by the sudden expansion of the brilliant plumage of the peacock, and the loud vibrations of his tail-feathers when he places himself in this strange attitude. A gorgeous spectacle suddenly presented, and so different from anything that is commonly seen, would overawe even the king of beasts. Similar effects in a weaker degree would be produced by the ruffled plumage of the turkey or the pheasant. It is worthy of remark, that in proportion to the brilliancy of the colors is the strength of the impression made upon the sight of the creature that threatens them. The tendency of wild animals to be frightened by such causes is shown by the terror produced in them by the sudden opening of an umbrella. But these brilliant plumes are confined to the larger species of the tribe. Quails, partridges, and grouse are generally colored like the ground, being of a speckled or brownish hue, and are distinguished with difficulty when sitting or standing among the berry-bushes or gleaning their repast in the cornfield. Too small to defend themselves so well as the larger species, their colors are adapted to protect them by concealment, and not by dazzling and alarming their foes.

# BIRDS OF THE GARDEN AND ORCHARD.

## III.

### THE ROBIN.

OUR American birds have not been celebrated in classic song. They are hardly well known even to our own people, and have not in general been exalted by praise above their real merits. We read, both in prose and verse, of the European Lark, the Linnet, and the Nightingale, and the English Robin Redbreast has been immortalized in song. But the American Robin is a bird of very different habits. Not much has been written about him as a songster, and he enjoys but little celebrity. He has never been puffed and overpraised, and though universally admired, the many who admire him are fearful all the while lest they are mistaken in their judgment and waste their admiration upon an object that is unworthy of it, — one whose true merits fall short of their own estimate. It is the same want of self-reliance affecting the generality of minds which often causes every man publicly to praise what each one privately condemns, thus creating a spurious public opinion.

I shall not ask pardon of those critics who are always canting about musical "power," and who would probably deny this gift to the Robin, because he cannot gobble like a turkey or squall like a cat, and because with his charming strains he does not mingle all sorts of discords and incongruous sounds, for assigning the Robin a very high rank as a singing-bird. Let them say, in the cant of modern criticism, that his performances cannot be great because they are faultless. It is enough for me that his mellow notes, heard at the earliest flush of dawn, in the busy hour of noon, or in the stillness of evening, come to the ear in a stream of unqualified melody, as if he had learned to sing from the beautiful Dryad who taught the Lark and the Nightingale. The Robin is surpassed by some other birds in certain qualities of song. The Mocking-Bird has more "power," the Red Thrush more variety, the Bobolink more animation; but there is no bird that has fewer faults than the Robin, or that would be more esteemed as a constant companion, — a vocalist for all hours, whose strains never tire and never offend.

There are thousands who admire the Mocking-Bird, because, after pouring forth a long-continued medley of disagreeable and ridiculous sounds, or a series of two or three notes, repeated more than a hundred times in uninterrupted and monotonous succession, he concludes with a single delightfully modulated strain. He often brings his tiresome extravaganzas to a magnificent climax of melody, and as often concludes an inimitable chant with a most contemptible bathos. But the notes of the Robin are all melodious, all delightful, loud without vociferation, mellow without monotony, fervent without ecstasy, and combining more of sweetness of tone, plaintiveness, cheerfulness, and propriety of utterance than the notes of any other bird.

The Robin is the Philomel of morning twilight in New England and in all the northeastern States of this continent. If his sweet notes were wanting, the mornings would be like a landscape without the rose, or a summer-evening sky without tints. He is the chief performer in the delightful anthem that welcomes the rising day. Of others the best are but accompaniments of more or less importance. Remove the Robin from this woodland orchestra, and it would be left without a *soprano*. Over all the northern parts of this continent, wherever there are human settlements, the Robins are numerous and familiar. There is not an orchard in New England, or in the British Provinces, that is not enlivened by several of these musicians. When we consider the millions thus distributed over this broad country, we can imagine the sublimity of that chorus which from the middle of April until the last of July daily ascends to heaven from the voices of these birds, not one male of which is silent from the earliest dawn until sunrise.

The Robin, when reared in confinement, is one of the most affectionate and interesting of birds. A neighbor and relative of mine kept one twenty years. He would leave his cage frequently, hop about the house and garden and return. He not only repeated his original notes, but several strains of artificial music. Though not prone to imitation, the Robin may be taught to imitate the notes of other birds. I heard a tamed Robin in Tennessee whistle "Over the Water to Charlie," without missing a note. Indeed, this bird is so tractable in his disposition and so intelligent, that I believe he might be taught to sing any simple melody.

But why should we set any value on his power of learning artificial music? Even if he should perform like a flautist, it would not enhance his value as a minstrel of the grove. We are concerned with the singing-birds only as they are in a state of nature and in their native fields and woods. It is the simplicity of their songs that constitutes their principal charm; and if the different warblers were so changed in their nature as to relinquish their wild notes and sing only tunes, we should listen to them with as much indifference as to the whistling of boys on the road.

## THE BALTIMORE ORIOLE.

About the middle of May, as soon as the cherry-trees are in blossom, and when the oak and the maple are beginning to unfold their plaited leaves, the loud and animated notes of the Golden Robin are first heard in New England. I have never known a bird of this species to arrive before that period. They seem to be governed by the supply of their insect-food, which probably becomes abundant at the same time with the flowering of the orchards. On their arrival they may be observed diligently hunting among the branches and foliage of the trees, making a particular examination of the blossoms for the flies and beetles that are lodged in them.

While the Oriole is thus employed in search of food, which he obtains almost exclusively from trees, he frequently utters his brief but loud and melodious notes. Of this species, the males arrive a few days before the females, and at first utter only a few call-notes, which on the arrival of their mates are lengthened into a song. This seldom consists of more than five or six notes, though the strain is sometimes immediately repeated. Almost all remarkable singing-birds give themselves up entirely to song on their musical occasions, and pay no regard to other demands upon their time until they have concluded. But the Golden Robin never relaxes from his industry, nor remains stationed upon the branch of a tree for the sole purpose of singing. He sings, like an industrious maid-of-all-work, only while employed in his sylvan occupations.

The Baltimore Oriole is said to inhabit North America from Canada to Mexico; but the species are most abundant in the northeastern parts of the continent, and a greater number of them breed in the New England States than either south or west of this section. They are also more numerous in villages and in the suburbs of cities than in the wilder regions where there is less tillage. Their peculiar manner of protecting their nests by hanging them from the spray of a tall elm or other lofty tree enables them to rear their young in security, even when surrounded by the dwellings of men. The only animals that are able to reach their nests are the smaller squirrels, which have been known to descend the long slender branches that sustain the nest, and to devour the eggs. This depredation I have never witnessed; but have seen the red squirrel descend in this manner upon the spray of an elm, and seize the chrysalis of a certain insect which was rolled up in a leaf.

The lively motions and general activity of the Golden Robin, no less than his song, render him interesting and attractive. He is remarkable for his vivacity, and his bright colors make all his movements conspicuous. His plumage needs no description, since every one is familiar with it, as its hues are seen like flashes of fire among the green foliage. Associated with these motions are his notes of anger and complaint, which have a peculiar vibratory sound, somewhat harsh, but not unmusical.

The Golden Robin is said to possess considerable power of musical imitation; but it may be observed that in all his attempts he gives the notes of those birds only whose voice resembles his own. Thus he often repeats the song of the Virginia Redbird. This I do not consider an imitation, but a mere change of his own melody in a slight degree. The few notes of his own song he utters frequently, and with great force and a fine modulation. Sometimes for several days he confines himself to a single strain, and then for about the same length of time he will adopt another. Sometimes he extends his few brief notes into a lengthened melody, and sings as in an ecstasy, like birds of the Finch tribe. Occasionally also he sings on the wing, not while hovering over one spot, but while flying from one tree to another. Such musical paroxysms are rare in his case, and seem to be caused by some momentary exultation.

The Golden Robin rears but one brood of young in New England, and his cheerful notes are discontinued soon after they have left their nest. The song of the old bird seems, after this event, hardly necessary as a call-note to the offspring, who keep up an incessant chirping from the moment of leaving their nest until they are able to accompany their parents to the woods. They probably retire to the forest for security, and vary their subsistence by searching for insects that occupy a wilder locality. It is remarkable that after an absence and silence of two or three weeks from the flight of their young the Golden Robins suddenly make their appearance once again for a few days, uttering the same merry notes with which they announced their arrival in May. But this renewal of their song is not continued many days. We seldom see them after the middle of August. They leave for their winter quarters early in autumn.

te-hoo, tee-hoo, te - oo, te-hoo, te-hoo, t - t - t - t, tee-hoo, te - oo.

### THE MEADOW-LARK.

This bird is no longer, as formerly, a Lark. Originally an Alauda, he has since been an Oriolus, an Icterus, a Cacicus, and a Sturnus. He has shuffled off all his former identities, and is now a Sturnella magna. I will not enter into a calculation of the metamorphoses he may yet undergo. By the magic charm of some inventor of another new nomenclature; by the ingenuity of some Kant in

Natural History, — if this science be doomed ever to suffer such a curse, when, by the use of new names for every thought of the human mind, we shall all be reduced to a sudden ignorance of everything we once knew, and rendered incapable of talking or writing without constant reference to a new dictionary of terms, — the Meadow-Lark may yet be discovered to be no bird at all, but a mere myth of the meadows.

The Meadow-Lark, though not the "Messenger of Morn" that "calls up the tuneful nations," and though perhaps not properly classed among our singing-birds, has a peculiar lisping note which is very agreeable, and not unlike some of the strains in the song of the English Wood-Lark, as I have heard them from a caged bird. Its notes are heard soon after those of the Robin, the earliest messenger of morn among our singing-birds. They are shrill, drawling, and plaintive, sometimes reminding me of the less musical notes of the Redwing and sometimes of the more musical and feeble song of the Green Warbler. Nuttall very aptly describes its notes by the syllables *et-see-dee-ah*, each one drawled out to a considerable length. These are repeated at all hours of the day; indeed, they are almost incessant, for hardly a minute passes when, if a pair of the birds are located in an adjoining field, you may not hear them. It is the constant repetition of their song that has led gunners to the discovery of the birds, which, if they had been silent, might have escaped notice.

That numerous class of men who would be more enraptured at the sight of "four-and-twenty blackbirds baked in a pie" than at the sound of their notes, though they equalled those of the Nightingale, — men who never look upon a bird save with the eyes and disposition of a prowling cat, and who display their knowledge of the feathered race chiefly at the gun-shops, — martial heroes among innocent songsters, — have not overlooked a bird so large and plump as the Meadow-Lark. Vain is its lisping and plaintive song; vain is the beauty displayed in its hovering and graceful flight, in its variegated plumage and its interesting ways! All these things serve but to render its species the more conspicuous mark for gunners, who have hunted them so incessantly that they are now as shy as the persecuted Crow, and as elusive a mark for the sportsman as a Loon.

Samuels says that "usually one bird of a flock is perched on a tree or a fence-post as a sentinel, and the moment a gunner approaches, the bird gives his alarm," when all the flock take wing. The Meadow-Lark is variegated above with different shades of yellow and brown; beneath, a lighter brown speckled with black. Its flight is very graceful, though not vigorous. The motions of its wings are rapid and intermittent, the slight pauses in their vibratory motions giving them a character quite unique.

## THE CEDAR-BIRD.

Little bird, that watchest the season of mellow fruits, and makest thy appearance like a guest who comes only on feast-days, and, like a truant urchin, takest the fair products of the garden without leave of the owner, saying not even a grace over thy meals like the Preacher, but silently taking thy fill, and then leaving without even a song of thankfulness, — still I will welcome thee to the festival of Nature, both for thy comely presence and thy cheerful and friendly habit with thy fellows.

The Cedar-Bird is not a songster. It seldom utters any note save the lisp that may always be heard when it is within sight. Dr. Brewer, who kept a wounded one in a cage, mentions that "beside its low, lisping call, this bird had a regular, faint attempt at a song of several low notes, uttered in so low a tone that it would be almost inaudible, even at a short distance. It became perfectly contented in confinement, and appeared fond of such members of the family as noticed it." He says of this species as proof of their devotion to one another and their offspring: "Once when one had been taken in a net spread over strawberries, its mate refused to leave it, suffered itself to be taken by the hand in its anxiety to free its mate, and, when set at liberty, would not leave until its mate had also been released and permitted to go with it."

According to Nuttall, during the mating season, they are always caressing each other like Turtle Doves. There is a manifestation of mutual fondness between these social birds. A friend assured him that he had seen one among a row of them seize an insect and offer it to its next neighbor, who passed it to the next, each politely declining the offer, until it had passed backwards and forwards several times.

The Cedar-Bird is not exclusively frugivorous. In the spring and early summer, before the berries are ripe, it feeds wholly upon insects and their larvæ. As a compensation for the mischief done by the bird and its fellows among the fruit-trees, they destroy vast numbers of canker-worms, taking them when they are very small and nestled in the flower-cup of the apple-tree. The excessive multiplication of the canker-worm seems a direct consequence of the proportional diminution of this and a few other valuable though mischievous species. Those cultivators who would gladly extirpate the boys as well as the birds, taking care to save boys enough to kill the birds, might, instead of persecuting the Cedar-Bird, find it more profitable in the end to pay a tax for its preservation.

This bird is very fond of the juniper. Its usual

abode is among the junipers. From these, when rambling in the woods, you will often start a flock; for they are easily alarmed on account of the pertinacity with which they have been hunted. It is seldom we see one bird of this species, without at least six or eight more in its company. Their habit of assembling in small flocks renders them more liable to be extirpated; for those who would grudge a charge of powder and shot for the flesh of a single bird are delighted to shoot into a flock, when perhaps six or eight little tender birds will fall to the ground.

The Cedar-Bird is remarkable for the elegance of its shape; and though the colors of its plumage are not brilliant, they are exceedingly fine and delicate. Its general color above is a reddish-brown, slightly tinged with olive; somewhat brighter on the breast, dark in the throat, tail tipped with yellow, forehead with a black line over the eyes, and little scarlet beads upon the outer wing-feathers, resembling dots of red sealing-wax.

## THE INDIGO-BIRD.

Some of the earliest nests I discovered in my boyhood were those of the Indigo-Bird, of which, for several successive years, there were two or three in a grove of young locust-trees near the building where I attended school. Hence I have always associated this bird with the locust-tree. Every one admires the beauty of the Indigo-Bird,— its plumage of dark-blue, with green reflections when in a certain light. Its color is not that of the Bluebird; but more nearly resembles a piece of indigo, being almost a blue-black. Though it never comes very near our windows, it does not appear to be shy, and it prefers the trees of our gardens and enclosures to those of the forest. When the breeding season is over, the old birds probably retire to the woods; for, after the young have taken flight, they are seldom seen.

I think Mr. Nuttall is incorrect in his description of the Indigo-Bird's song. It certainly has not that variety and pathos which he ascribes to it. The song is rather a lively see-saw without expressing even animation. It ought not to be considered plaintive. His notes are sharp, not unlike those parts of the Canary's song which are disagreeable. I allude to the *sip, sip, sip, sip,* which the Canary intersperses with his more musical and rolling notes. The whole song of the Indigo-Bird is but a repetition of the *sip, sip,* of the Canary, modified by the addition of another note, like *sip-see, sip-see, sip-see, sip-see,* repeated four or five times very moderately, with a few unimportant intervening notes. Neither has the song of the Indigo-Bird so much rapidity as Nuttall ascribes to it. His notes, though not slow, are but little more rapid than those of the Robin. He has the merit, however, of being one of the few of our birds that sing persistently at noonday.

### THE SUMMER YELLOW-BIRD.

There is no common feature in our New England domestic landscape more remarkable than the frequent rows of willows which have at different times been planted by the sides of roads where they pass over wet meadows. The air is never sweeter, not even in a grove of lindens, than the vernal breezes that are constantly playing among the willows, when they are hung with golden aments, and swarming with bees and butterflies. Here, flitting among the soft foliage of these trees after the middle of May, you will never fail to meet the little Summer Yellow-Bird, whose plumage is so near the color of the willow-blossoms that they almost conceal it from observation.

The Summer Yellow-Bird is one of that incomparable tribe of warblers, comprehended under the general name of sylvians, that frequents familiar places. His plumage is not a bright yellow, but faintly streaked with olive on the back and wings. He feeds entirely on insects, and is frequently seen in gardens among the cherry-trees and currant-bushes in search of them. The birds of this species are not shy; and I have observed the same confiding docility in other small birds which are not persecuted. The note of the male is remarkable only for its sweetness. It is too brief and shrill to attract attention, except by giving notice of the cheerful presence of the bird. He is so familiar as frequently to come up close to our windows when a tree is near, peeping in upon us as if to watch our motions.

There is nothing in his general habits to render him conspicuous; and little is said about him, because he is quiet and unobtrusive. But were his whole species banished from our land, he would be missed as we should miss the little cinqfoil from our green hillsides, which it sprinkles with its modest and familiar flowers, though it attracts no admiration. The Summer Yellow-Bird, like this little flower, dwells sweetly among the willows and cherry-trees, seen by all, and loved for its unpretending beauty, its cheerful note, and its innocent habits.

Dr. Brewer mentions the Summer Yellow-Bird as one of the few species that refuses to hatch the egg of the Cowbird. If this bird should drop one of its eggs into her nest, she builds up the walls and then covers the spurious egg with a thick coating of fresh materials. He mentions one remarkable case that happened in his own garden. The Yellow-Bird had already built a new nest over one Cowbird's egg. Another was deposited in the new nest, and she built over that. She had finally made a nest with three stories, the last one containing only the Yellow-Bird's eggs. This fact and others of a similar kind, related by ornithologists, indicate an unusual share of intelligence in this species. Dr. Brewer also mentions an anecdote related to him by a friend. A pair of Yellow-Birds had built their nest in a low bush, and filled it with eggs, when a storm partly overturned it. They abandoned it and built another in the same bush, and the female laid her eggs and sat upon them. "The narrator then restored the first nest to an upright position and securely fastened it." The male bird immediately sat upon the eggs in this nest, while the female sat upon the other. In this way each one hatched, fed, and reared its separate family.

# THE FIELD AND THE GARDEN.

It must have been observed by every careful student of nature that our walks in the field and in the garden are not attended by the same sensations. Indeed, they always remind me of prose and verse, the one marked by uniformity, the other by variety. The words and images of prose are more ample and free, those of verse more select and condensed. We look for assorted profusion in the garden, for scattered multiplicity in the field. We can sustain our interest a longer time when rambling over the fields of prose; but the luxury of a few moments is greater when traversing the garden-walks of a short poem. We see more beauty, more splendor, more that gratifies the sense, in the garden; we discover more of the picturesque, more sublimity, more that excites the imagination, in the field. But the dreary monotony and artificial grandeur of a widely extended landscape garden must be as tiresome as a long poem; its serpentine paths, rustic devices, and shallow imitations of nature's wildness failing in their intentions, as the affected ruggedness and hobbling of the verse and the frequent episodes of a long poem are but a mockery of the freedom of prose.

People who have been confined a great part of their life to the town know very little of flowers, except as the ornaments of a garden, and have admired them chiefly as objects of art. Florists' flowers are generally deprived of some of their specific characters: stamens are transformed into petals, as in roses; wheel-shaped flowers in the margin take the place of bell-shaped flowers in the centre, as in the snowball; or the florets of the disk are furnished with petals, as in the dahlia, and become in each case a "double flower." By this transformation they are rendered more valuable for bouquets and floral exhibitions, and are more admirable as ornaments of the parterre. They have become more marketable, but less poetical; they are more the delight of the flower-girl, but they are prized in a less degree by the botanist and the poet, who prefer the objects of nature unsophisticated by art.

The field-flowers are praised by the poet Campbell, because they waft him to bygone summers, to birchen glades and Highland mountains, to the shores of lakes and their little islands; because they are associated with the notes of birds and the voices of streams. While admitting that they are eclipsed by the flowers of the garden, he gives these wildings of nature his preference, because they are allied with more pleasant memories and affections. He would cherish them that they may enliven his declining years with the sensations of youth, and hopes they may grow upon his tomb. The simple flowers of the garden, however, which have not been greatly modified by culture and retain their original characters, claim no less attention than we bestow upon the flowers of the field. The most ancient and common of these have acquired the greatest share of our affection, because they were our earliest friends. Such are the primrose, the pansy, the narcissus, the tulip, the lily of the valley,— perfectly primitive in its character, — and above all, the white lily and the rose. We have become acquainted with these flowers, not only from our early intercourse with them in the garden, but from the frequent allusions to them in the poetry of all ages, and in Holy Writ. But they are not the favorites of florists. Fashion, who always impudently interferes with our tastes and our pleasures, has not failed to intermeddle with the flower-garden, and has often stamped a false value upon certain flowers of inferior beauty compared with others of a more simple habit and deportment. We who have not been compelled to wear the yoke of this tyranny will continue to admire those which have been sanctified to our imagination by the poets of nature.

Many of our common garden-flowers are closely interwoven with the fabric of English literature; and the frequent mention of them by the early poets, who treated them more in detail than their successors, has invested them with charms which are derived from their descriptions and the imagery that accompanies them. Others are commended to us by the memories of childhood, and by their frequency in the gardens of rustic cottages in the country. Such are the marigold, the larkspur, the morning-glory, the iris, the crocus, and the snowdrop. How vividly are the early scenes and events of our life called up by these simple flowers, and how greatly do they contribute to the cheerfulness and sacredness of the grounds they occupy! Coming generations will be affected with less emotion by these particular flowers, because their childhood will make friendships with others that have taken their places. But I am persuaded that the introduction of such multitudinous species in our gardens is fatal to the poetic interest that might be felt in a smaller number. A few flowers take a stronger hold of our affections and our imaginations than a multitude. Thus people who live in retirement, with a small circle of friends, are more devoutly attached to them than others who have very many, whom they constantly meet in the social intercourse of fashion.

I will confess that I am not an admirer of floral exhibitions. I am offended when I see flowers degraded to a level with ribbons, laces, and jewelry, and prized according to some property that is appreciable only by a connoisseur. I am aware that such exhibitions are attended with certain public advantages, and contribute an innocent amusement to the inhabitants of towns and cities. But I should be more interested in looking over the dried specimens of some rustic botanist in the country than in viewing the most splendid assortment of show-flowers; and feel more respect for the zeal of a true lover of nature, who traverses the continent in quest of an unknown species, than for the ambition of a florist, who experiments half his lifetime to add one new tint to a dahlia.

I was invited some time since by an old lady of my acquaintance to visit her garden and see her flowers, of which she had gathered together a miscellaneous assemblage that reminded me of those we sometimes meet in a little opening in the woods. She was one who valued plants as the works of nature, not as the toys of ambition, and who held them all sacred as gifts of Providence. Every species was highly prized by her, and she had collected all such as her means enabled her to obtain, and planted them in her garden. This little enclosure I found to be stored with many plants which have been naturalized on our soil, and from time immemorial have been known and loved by the inhabitants both of England and America. Many of these were common in our gardens thirty years ago. Among them were several cordial and medicinal herbs, such as wormwood, balm, horehound, southernwood, basil, and thyme, growing side by side with pinks, jasmines, and primulus. She expatiated on the uses of these and the beauties of those; but the principal objects of her admiration were some noble sunflowers, that maintained a sort of kingly presence among the inhabitants of her garden.

Not being affected by any prejudice against sunflowers, I sympathized with her admiration, and praised them heartily without saying a word more than I felt. They were dotted about her grounds with great irregularity, not because the old lady had any of the prevailing affectation for what is termed picturesque arrangement, but wherever a seed had come up, there she allowed it to grow without molestation. There was an air of rustic cheerfulness about these sunflowers that captivated my sight, and made me at the time a true convert to the views of my entertainer. This celebrated flower, which was dedicated to the sun, because it was made in the image of that deity, — the flower which was produced by the transformation of Clytie, and, still retaining her passion, is supposed to turn itself constantly toward his beams, — had found a modern admirer in my hostess. Though its colors are neither various nor beautiful, there is a halo of divinity in the border of petals surrounding the disk of the flower, and a look that reminds me of those charitable and honest people who live to do good. We shall perceive this analogy when we consider that the sunflower possesses many economical properties, and that, after the beauty of its prime is faded, it scatters abroad its seeds, and supplies a repast for many famishing birds. The good dame appreciated these frugal habits in her sunflowers, and fed her poultry in the autumn with their seeds.

While commenting on the beauties of the various occupants of her garden, she made an apology for the weeds which had overgrown and concealed many of her favorite flowers; her duties as a housekeeper had not left her time enough to be a good supervisor of her plants. I remarked that weeds are an important addition to a flower-garden; that they cause it to resemble the wilds of Nature, who is not careful to destroy weeds, but seems as desirous to protect them as the most beautiful lilies or daisies. It is pleasant when strolling in a garden to feel as if we were making discoveries, by gaining perhaps the first sight of a little blossom half hidden by some overtopping weed. She did not quite comprehend my philosophy, and thought it preferable that the beauties of the garden should be the most conspicuous objects. I replied that many of her weeds were as beautiful as her flowers; that the Roman wormwood, for example, generally despised, was nothing less than the Ambrosia, which was served with nectar at the feasts of the gods; it is like a tree in its manner of branching, and bears a leaf like that of a fern, — the proudest of all plants in the structure of its foliage.

On our way through the garden-path a large burdock in an angle of the fence obtruded itself upon our sight, covered with a splendid array of purple globular flowers. The burdock, she said, was allowed to occupy this obscure nook for the benefit of its seeds, which, if made into a tea, are a valuable remedy for weak nerves; and she often steeped its roots with certain aromatic herbs, to add a tonic bitter to her "diet drink." I added that it was once highly prized as a medicinal herb, and that, setting aside the beauty of its flowers, I should cherish this particular one for the protection it afforded to a little creeping plant then luxuriating in its shade. This little creeper was the gill; a very pretty labiate, displaying its blue and purple flowers in whorls, and the stem with anthers that meet and form a cross, and adorned with heart-shaped leaves very neatly corrugated. This plant had gained my admiration very early in life, among the weeds in my own garden, and on account of its delicate beauty I could not treat it as an outcast.

Among other curiosities of her garden, included in the denomination of weeds, was a delicate euphorbia, a flat spreading plant, lying so close upon the ground that it could hardly be touched by the foot that was placed upon it. It grew in the garden walk, forming circular patches, and covered with minute round leaves, having a purple spot in their centre, and bearing in their axils a little white flower. This plant had not attracted her attention, and she seemed pleased at having made so rare a discovery among her weeds. On the other hand, she had not failed to observe a beautiful sandwort, one of the most delicate of nature's productions, with a profusion of small pink flowers upon stalks and leaves as fine as moss. This had planted itself on a rude terrace near the walls of her cottage, where the sandy soil would not permit the growth of more luxuriant plants that would overshadow and destroy it. She seemed to admire this little weed as much as her sunflowers, and had taken notice of the fine hues of its corolla, its branching stems, and its leaves terminating in fine bristles. Before we separated I remarked that her weeds required no apology, for after all they were not so numerous as to hold any more than their rightful share of the soil. I confessed that in the neglected parts of her garden I had obtained as much satisfaction as if it were a proud parterre; that there might be an excess of beauty and elegance in a garden as well as in a dwelling-house. My visit had been an exceedingly pleasant one to me; and I cared no more to see a garden where everything is kept in as nice a trim as the bald pate of a Chinaman, than to look at the pictures in a barber's shop.

I soon afterwards entered the grounds of an amateur florist, who showed me a fine array of the most recently imported florists' flowers. He discoursed eloquently on the superiority of certain improved dahlias, compared with other similar varieties that might seem identical to one who is not a connoisseur. He was particularly pleased with some beds of hollyhocks that displayed a great variety of colors and shades, which he had combined so as to produce a beautiful harmonic effect that reminded me of the colors of the rainbow. I could not help saying that I admired the splendor of this exhibition, and the ingenuity required for its arrangement; but I did not praise it sufficiently to gratify his ambition, and he expressed his surprise at my want of enthusiasm. I soon perceived that he was, in the most approved sense, a man of taste and of "æsthetic culture"; that he had a keen eye for any improvement in a flower as manifested in a new combination of hues or rare development of form, and great skill in the arrangement of his borders. More than all, he was so much of a scientific botanist that I was instructed by his discourse no less than I had been delighted by my interview with his humble neighbor.

He alluded to my visit in the old lady's garden, and spoke in a comical humor of her sunflowers and her admiration of them. I replied that whole nations had worshipped the sun; and why should not our pious friend worship the sunflower, which is typical of that luminary? This religion of hers was a proof of her admiration of greatness, in which she resembled the rest of the world. The public has never ceased to admire big trees and mammoth squashes; and a great sunflower seems to me as worthy of our idolatry as a great water-lily. I confessed that I could join heartily in the respect she paid even to her burdocks, that bear a profusion of flowers, consisting of little globular beads of the most exquisite finish, with tufts of rose-colored fringe, each one a gem fit to adorn the bosom of a sylph. These plants are also of a giant size, with a leaf as large as that of a fan-palm. I added that I felt a homely regard for flowers, not in proportion as they were "far-fetched and dear-bought," but as they are adapted to certain important ends connected with our happiness, independent of our ambition. I left him in a state of surprise at my avowal of so many heresies which he thought disproved my sincerity. But I am not able to perceive the superiority of his taste compared with that of my female friend. I cannot understand why mere splendor is a thing to be admired, or simplicity a thing to be ridiculed. A true painter sees more to delight him in a laborer's cottage guarded by an old apple-tree, than in a palace surrounded by works of sculpture and shaded by cedars of Lebanon.

There is an inclination among men to carry their social prejudices into their observations of nature, to make price a criterion of beauty as well as of value, and to qualify their admiration of both scenes and flowers by their ideas of the expense which has been laid out upon them. This is the way to annihilate everything sacred and poetical in the character of flowers and landscape, and to degrade nature below art, or, rather, I should say, below fashion. The simple-hearted woman who cherishes with fondness a lilac-tree that bore flowers for her when she was a girl, manifests a sentiment that is entitled to respect, and her affection for it is a genuine theme for poetry. He who despises her attachment because her lilac-tree is out of date as a thing of fashion, and has lost its value in the flower-market, is himself the proper subject of satire. Let us save these fair objects of the field and the garden from being appraised like millinery goods!

# MAY.

THE spring in New England does not, like the same season in high northern latitudes, awake suddenly into verdure out of the bosom of the snows. It lingers along for more than two months from its commencement, like that long twilight of purple and crimson that leads up the mornings in summer. It is a pleasant, though sometimes weary prolongation of the season of hopes and promises, frequently interrupted by short periods of wintry gloom. The constant lingering delay of nature in the opening of the flowers and the leafing of the trees affords us something like an extension of the dayspring of life and its joyful anticipations. As we ramble through rustic paths and narrow lanes and over meadows still dank and sere, the very tardiness with which the little starry blossoms peep out of its darkness, and with which the wreath of verdure is slowly drawn over the plains, gives us opportunity to watch them and become acquainted with their beauty, before they are lost in the crowd that will soon appear.

Our ideas of May, being derived, in part, from the descriptions of English poets and rural authors, abound in many pleasant fallacies. There are no seas of waving grass and bending grain in the May of New England. Nature is not yet clothed in the fulness of her beauty; but in many respects she is lovelier than she will ever be in the future. Her very imperfections are charming, inasmuch as they are the budding of perfection, and afford us the agreeable sentiment of beauty joined with that of progression. It is this thought that renders a young girl more lovely and interesting with her unfinished graces than when she has attained the completion of her charms. The bud, if not more beautiful, is more poetical than the flower, as hope is more delightful than fruition.

The ever-changing aspects of the woods are sources of continual pleasure to the observer of nature, and have in all ages afforded themes for the poet and subjects for the painter. Of all these phases the one that is presented to the eye in May is by far the most delightful, on account of the infinite variety of tints and shades in the budding and expanding leaves and blossoms, and the poetic relations of their appearance at this time to the agreeable sentiment of progression. The unfolding leaves and ripening hues of the landscape require no forced effort of ingenuity to make apparent their analogy to the period of youth and season of anticipation; neither are the fading tints of autumn any less suggestive of life's decline. There are not many, however, who would not prefer the lightness of heart that is produced by these emblems of progression, and these signals of the reviving year, to the more poetic sentiment of melancholy inspired by the scenes of autumn.

It is pleasant at this time to watch the progress of vegetation, from the earliest greenness of the meadow, and the first sprouting of the herbs, unfolding of the leaves, and opening of the buds, until every herb, tree, and flower has expanded and brightened into the full radiance of summer. While the earth displays only a few occasional stripes of verdure along the borders of the shallow pools and rivulets, and on the hillsides, where they are watered by oozing fountains just beneath the surface, we may observe the beautiful drapery of the tasselled trees and shrubs, varying in color from a light yellow to a dark orange or brown, and robing the landscape in a flowery splendor that forms a striking contrast with the general nakedness of the plain. As the hues of this drapery fade by the withering of the catkins, the leaf-buds of the trees gradually cast off their scaly coverings, in which the infant bud has been cradled during the winter; and the tender fan-shaped leaves, in plaited folds of different hues, come forth in millions, and yield to the forest a golden and ruddy splendor, like the tints of the clouds that curtain the summer horizon.

There is an indefinable beauty in the infinitely varied hues of the foliage at this time, yet they are far from being the most attractive spectacle of the season. While the trees are unfolding their leaves, the earth is daily becoming greener with every nightfall of dew, and thousands of flowers awake into life with every morning sun. At first a few violets appear on the hillsides, increasing daily in numbers and brightness, until they are more numerous than the stars of heaven; then a single dandelion, which is the harbinger of millions in less than a week. All these gradually multiply and bring along in their rear a countless troop of anemones, saxifrages, geraniums, buttercups, and columbines, until the landscape is draped with the universal wreath of spring.

May opens with a few blossoms of the coltsfoot, the liverwort, the buckbean, and the wood-anemone, and a multitude of blue violets of a humble species, such as we see upon the grassy mounds in our old country graveyards, are scattered over the southern slopes of the pastures. After May-day, every morning sun is greeted by a fresh troop of these little fairy visitants, until every nook sparkles with them, and every pathway is embroidered with them. At an early period the green pastures are so full of dandelions and buttercups that they seem to be smiling upon us from every knoll. Children are always delighted with these flowers, and our eyes, as they wander over the village outskirts, will rest upon hundreds of young children, on a sunny afternoon, who have left their active sports to gather them and weave them into garlands, or use them as talismans with which they have associated many interesting conceits. Soon after this the fields appear in the fulness of their glory. Wild geraniums in the borders of the woods and copses, white and yellow violets, ginsengs, bellworts, silverweeds, and cinquefoils bring up the rear in the procession of May. During all this time the flowers of the houstonia, which have been very aptly chosen as the symbols of innocence, beginning in the latter part of April with a few scanty blossoms, grow every day more and more abundant, until their myriads resemble a thin but interminable wreath of snowflakes, distributed over the hills and pastures.

If we now look upon the forest, we shall observe a manifest connection between the tints of the half-developed spring foliage and those observed in the decline of the year. The leaves of nearly all the trees and shrubs that are brightly colored in autumn present a similar variety of tints in their plaited foliage in May. It is these different tendencies of all the various species that afford the woods their principal charm during this month.' It seems, indeed, to be the design of nature to foreshow, in the infancy of the leaves, some of those habits that mark both their maturity and their decline, by giving them a faint shade of the colors that distinguish them in the autumn.

Though we cannot find in May those brilliant colors among the leaves of the forest-trees which are the crowning glory of autumn, yet the present month is more abundant in contrasts than any other period, and these increase in beauty and variety until about the first of June. In early May, set apart from the general nakedness of the woods, may be seen here and there a clump or a row of willows full of bright yellow aments, maples with buds, blossoms, and foliage of crimson, and interspersed among them junipers, hemlocks, and other evergreens, that stand out from their assemblages like natives of another clime. As the month advances, while these contrasts remain, new ones are daily appearing as one tree after another comes into flower, each exhibiting a tint peculiar not only to the species, but often to the individual and the situation, until hardly two trees in the whole wood are quite alike in color.

As the foliage ripens, the different shades of green become more thoroughly blended into a single uniform tint. But ere the process is completed the fruit-trees open their blossoms and bring a new spectacle of contrasts into view. Peach-trees, with their pale crimson flowers, that appear before the leaves, and stand in flaming rows along the fences, like burning bushes; pear-trees, with corols perfectly white, internally fringed with brown anthers, like long dark eyelashes, that give them almost the countenance of life; cherry-trees, with their white flowers enveloped in tufts of foliage, occupied by the oriole and the linnet; and apple-trees, with flowers of every shade between a bright crimson or purple and a pure white, — all come forth one after another to welcome the birthday of June.

During the last week in May, were you to stand on an eminence that commands an extensive view of the country, you would be persuaded that the prospect is far more magnificent than at midsummer. At this time you look not upon individuals but upon groups. Before you lies an ample meadow, nearly destitute of trees save a few elms standing in equal majesty and beauty, combining in their forms the gracefulness of the palm with the grandeur of the oak; here and there a clump of pines, and long rows of birches, willows, and alders bordering the streams that glide along the valley, and displaying every shade of green in their foliage. In all parts of the prospect, separated by square fields of tillage of lighter and darker verdure, according to the nature of their crops, you behold numerous orchards, — some, on the hillside, receiving the direct beams of the sun ; others, on level ground, exhibiting their shady rows with their flowers just in that state of advancement that serves to show the budding trees, which are red and purple, in beautiful opposition to the full-blown trees, which are white. Such spectacles of flowering orchards are seen in all parts of the country, as far as the eye can reach along the thinly inhabited roadsides and farms.

The air at this time is scented with every variety of perfumes, and every new path in our rambling brings us into a new atmosphere as well as a new prospect. It is during the prevalence of a still south-wind that the herbs and flowers exhale their most agreeable odors. Plants generate more fragrance in a warm air; and if the wind is still and moist, the odors, as they escape, are not so widely dissipated, being retained near the ground by mixing with the dampness of the atmosphere. Hence the time when the breath of flowers is sweetest is during a calm, when the weather is rather sultry, and while the sunbeams are tinged with a purple and ruddy glow by shining through an almost invisible haze. A blind man might then determine, by the perfumes of the air, as he was led over the country, whether he was in meadow or upland, and distinguish the character of the vegetation.

Now let the dweller in the city who, though abounding in riches, sighs for that contentment which his wealth has not procured, come forth from the dust and confinement of the town and pay a short visit to Nature in the country. Let him come in the afternoon, when the declining sun casts a beautiful sheen upon the tender leaves of the forest, and while thousands of birds are chanting, in full chorus, from an overflow of those delightful sensations that fill the hearts of all creatures who worship Nature in her own temples and do obedience to her beneficent laws. I would lead him to a commanding view of this lovely prospect, that he may gaze awhile upon those scenes which he has so often admired on the canvas of the painter, displayed here in all their living beauty. While the gales are wafting to his senses the fragrance of the surrounding groves and orchards, and the notes of the birds are echoing all around in harmonious confusion, I would point out to him the neat little cottages which are dotted about like palaces of content in all parts of the landscape. I would direct his attention to the happy laborers in the field, and the neatly dressed, smiling, ruddy, and playful children in their green and flowery enclosures and before the open doors of the cottages. I would then ask him if he is still ignorant of the cause of his own unhappiness, or of the abundant sources of enjoyment which Nature freely offers for the participation of all her creatures.

# THE ANTHEM OF MORN.

NATURE, for the delight of waking eyes, has arrayed the morning heavens in the loveliest hues of beauty. Fearing to dazzle by an excess of light, she first announces day by a faint and glimmering twilight, then sheds a purple tint over the brows of the rising morn, and infuses a transparent ruddiness throughout the atmosphere. As daylight widens, successive groups of mottled and rosy-bosomed clouds assemble on the gilded sphere, and, crowned with wreaths of fickle rainbows, spread a mirrored flush over hill, grove, and lake, and every village spire is burnished with their splendor. At length, through crimsoned vapors, we behold the sun's broad disk, rising with a countenance so serene that every eye may view him ere he arrays himself in his meridian brightness. Not many people who live in towns are aware of the pleasure attending a ramble near the woods and orchards at daybreak in the early part of summer. The drowsiness we feel on rising from our beds is gradually dispelled by the clear and healthful breezes of early day, and we soon experience an unusual amount of vigor and elasticity. Nature has so ordered her bounties and her blessings as to cause the hour which is consecrated to health to be attended with the greatest number of charms for all the senses; and to make all hearts enamored of the morning, she has environed it with everything, in heaven and on earth, that is delightful to the eye or to the ear, or capable of inspiring some agreeable sentiment.

During the night the stillness of all things is the circumstance that most powerfully attracts our notice, rendering us peculiarly sensitive to every accidental sound that meets the ear. In the morning, at this time of year, on the contrary, we are overwhelmed by the vocal and multitudinous chorus of the feathered tribe. If you would hear the commencement of this grand anthem of nature, you must rise at the very first appearance of dawn, before the twilight has formed a complete semicircle above the eastern porch of heaven. The first note that proceeds from the little warbling host is the shrill chirp of the hair-bird, — occasionally vocal at all hours on a warm summer night. This strain, which is a continued trilling sound, is repeated with diminishing intervals, until it becomes almost incessant. But ere the hair-bird has uttered many notes a single robin begins to warble from a neighboring orchard, soon followed by others, increasing in numbers until, by the time the eastern sky is flushed with crimson, every male robin in the country round is singing with fervor.

It would be difficult to note the exact order in which the different birds successively begin their parts in this performance; but the bluebird, whose song is only a short mellow warble, is heard nearly at the same time with the robin, and the song-sparrow joins them soon after with his brief but finely modulated strain. The different species follow rapidly, one after another, in the chorus, until the whole welkin rings with their matin hymn of gladness. I have often wondered that the almost simultaneous utterance of so many different notes should produce no discords, and that they should result in such complete harmony. In this multitudinous confusion of voices, no two notes are confounded, and none has sufficient duration to grate harshly with a dissimilar sound. Though each performer sings only a few strains and then makes a pause, the whole multitude succeed one another with such rapidity that we hear an uninterrupted flow of music until the broad light of day invites them to other employments.

When there is just light enough to distinguish the birds, we may observe, here and there, a single swallow perched on the roof of a barn or shed, repeating two twittering notes incessantly, with a quick turn and a hop at every note he utters. It would seem to be the design of the bird to attract the attention of his mate, and this motion seems to be made to assist her in discovering his position. As soon as the light has tempted him to fly abroad, this twittering strain is uttered more like a continued song, as he flits rapidly through the air. But at this later moment the purple martins have commenced their more melodious chattering, so loudly as to attract for a while the most of our attention. There is not a sound in nature so cheering and animating as the song of the purple martin, and none so well calculated to drive away melancholy. Though not one of the earliest voices to be heard, the chorus is perceptibly more loud and effective when this bird has united with the choir.

When the flush of morning has brightened into vermilion, and the place from which the sun is soon to emerge has attained a dazzling brilliancy, the robins are already less tuneful. They are now becoming busy in collecting food for their morning repast, and one by one they leave the trees, and may be seen hopping upon the tilled ground, in quest of the worms and insects that have crept out during the night from their subterranean retreats. But as the robins grow silent, the bobolinks begin their vocal revelries; and to a fanciful mind it might seem that the robins had gradually resigned their part in the performance to the bobolinks, not one of which is heard until some of the former have concluded their songs. The little hair-bird still continues his almost incessant chirping, the first to begin and the last to quit the performance. Though the voice of this bird is not very sweetly modulated, it blends harmoniously with the notes of other birds, and greatly increases the charming effect of the combination.

It would be tedious to name all the birds that take part in this chorus; but we must not omit the pewee, with his melancholy ditty, occasionally heard like a short minor strain in an oratorio; nor the oriole, who is really one of the chief performers, and who, as his bright plumage flashes upon the sight, warbles forth a few notes so clear and mellow as to be heard above every other sound. Adding a pleasing variety to all this harmony, the lisping notes of the meadow-lark, uttered in a shrill tone, and with a peculiarly pensive modulation, are plainly audible, with short rests between each repetition.

There is a little brown sparrow, resembling the hair-bird, save a general tint of russet in his plumage, that may be heard distinctly among the warbling host. He is rarely seen in cultivated grounds, but frequents the wild pastures, and is the bird that warbles so sweetly at midsummer, when the whortleberries are ripe, and the fields are beautifully spangled with red lilies. There is no confusion in the notes of his song, which consists of one syllable rapidly repeated, but increasing in rapidity and rising to a higher key towards the conclusion. He sometimes prolongs his strain, when his notes are observed to rise and fall in succession. These plaintive and expressive notes are very loud and constantly uttered, during the hour that precedes the rising of the sun. A dozen warblers of this species, singing in concert, and distributed in different parts of the field, form, perhaps, the most delightful part of the woodland oratorio to which we have listened.

As the woods are the residence of a tribe of musicians that differ from those we hear in the open fields and orchards, we must spend a morning in each of these situations, to obtain a hearing of all the songsters of daybreak. For this reason I have said nothing of the thrushes, that sing chiefly in the woods and solitary pastures, and are commonly more musical in the early evening than in the morning. I have confined my remarks chiefly to those birds that frequent the orchards and gardens, and dwell familiarly near the habitations of men.

At sunrise hardly a robin can be heard in the whole neighborhood, and the character of the performance has completely changed during the last half-hour. The first part was more melodious and tranquillizing, the last is more brilliant and animating. The grass-finches, the vireos, the wrens, and the linnets have joined their voices to the chorus, and the bobolinks are loudest in their song. But the notes of birds in general are not so incessant as before sunrise. One by one they discontinue their lays, until at high noon the bobolink and the warbling flycatcher are almost the only vocalists to be heard in the fields.

Among the agreeable accompaniments of a summer morning walk are the odors from the woods, the herbage, and the flowers. At no other hour of the day is the atmosphere so fragrant with their emanations. The blossoms of almost every species of plant are just unfolding their petals, after the sleep of night, and their various offerings of incense are now poured out at the ruddy shrine of morning. The objects of sight and sound are generally the most expressive in a description of nature, because seeing and hearing are the intellectual senses. But the perfumes that abound in different situations are hardly less suggestive than sights and sounds. Let a person who has always been familiar with green fields and babbling brooks, and who has suddenly become blind, be led out under the open sky, and how would the various perfumes from vegetation suggest to him all the individual scenes and objects that have been imprinted on his memory!

There is a peculiar feeling of hope and cheerfulness that attends a summer morning walk, and spreads its happy influence over all the rest of the day. The pleasant stillness, apart from the stirring population; the amber glow of heaven that beams from underneath successive arches of crimson and vermilion, constantly widening and brightening into the full glory of sunrise; the melodious concert of warblers from every bush and tree, constantly changing its character by the silence of the first performers and the joining of new voices, — all conspire to render the brief period from dawn to sunrise a consecrated hour, and to sanctify it to every one's memory. I am inclined to attribute the healthfulness of early rising to these circumstances rather than to the doubtful salubrity of the dewy atmosphere of morn. The exercise of the senses while watching the beautiful gradations of colors, through which the rising luminary passes ere his full form appears in sight, is attended with emotions like those which might be supposed to attend us at the actual opening of the gates of Paradise. We return home after this ramble warmed by new love for the beautiful objects of nature, and with all our feelings so harmonized by the sweet influences of morn as to find increased delight in the performance of our duties and the exercise of our affections.

# BIRDS OF THE PASTURE AND FOREST.

## I.

HE who has always lived in the city or its suburbs, who has seldom visited the interior except for purposes of trade, and whose walks have not often extended beyond those roads which are bordered on each side by shops and dwelling-houses, may never have heard some of our most remarkable songsters. These are the birds of the pasture and forest, those shy, melodious warblers who sing only in the ancient haunts of the Dryads. These birds have not multiplied like the familiar birds in the same proportion with the increase of human population and the extension of agriculture. Though they do not shun mankind, they keep aloof from villages, living chiefly in the deep wood or on the edge of the forest and in the bushy pasture.

There is a peculiar wildness in the songs of this class of birds that awakens a delightful mood of mind, similar to that which is excited by reading the figurative lyrics of a romantic age. This feeling is undoubtedly, to a certain extent, the effect of association. Having always heard their notes in wild and wooded places, they never fail to bring this kind of scenery vividly before the imagination, and their voices are like the sounds of mountain streams. It is certain that the notes of the solitary birds do not affect us like those of the Robin and the Linnet; and their influence is the same, whether it be attributable to some intrinsic quality or to association, which is indeed the source of some of the most delightful emotions of the human soul.

Nature has made all her scenes and the sights and sounds that accompany them more lovely by causing them to be respectively suggestive of a peculiar class of sensations. The birds of the pasture and forest are not frequent enough in cultivated places to be associated with our homes and our gardens. Nature has confined certain species of birds and animals to particular localities, and thereby gives a poetic or picturesque attraction to their features. There are certain flowers that cannot be cultivated in a garden, as if they were designed for the exclusive adornment of those secluded arbors which the spade and the plough have never profaned. Here flowers grow which are too holy for culture, and birds sing whose voices were never heard in the cage of the voluptuary, and whose tones inspire us with a sense of freedom known only to those who often retire from the world to live in religious communion with nature.

### THE SWAMP-SPARROW.

There is a little Sparrow whose notes I often hear about the shores of unfrequented ponds, and from their untrodden islets covered with button-bush and sweet gale, and never in any other situations. The sound of his voice always enhances the sensation of rude solitude with which I look upon this primitive scenery. We often see him perched upon the branch of a dead tree that stands in the water, a few rods from the shore, apparently watching our angling operations from his leafless perch, where he sings so sweetly that the very desolation of the scene borrows a charm from his voice that renders every object delightful.

This little solitary warbler is the Swamp-Sparrow. He bears some resemblance to the Song-Sparrow, but he is without that bird's charming variety of modulation. His notes have a peculiar liquid tone, and sometimes resemble the rapid dropping of water by the single drop into a wooden cistern which is half full. They may be compared to the trilling of the Hair-Bird, a kindred species, less rapidly uttered, and upon a lower key. If their notes are not plaintive, as Nuttall considered them, they produce very vividly a sensation of solitude, that tempts you to listen long and patiently, as to a sweet strain in some rude ballad music.

### THE WOOD-SPARROW.

When the flowers of early summer are gone, and the graceful neottia is seen in the meadows, extending its spiral clusters among the nodding grasses; when the purple orchis is glowing in the wet grounds, and the roadsides are gleaming with the yellow blossoms of the hypericum, the merry voice of the Bobolink has ceased and many other familiar birds have become silent. At this time, if we stroll away from the farm and the orchard into more retired and wooded haunts, we may hear at all hours and at frequent intervals the pensive and melodious notes of the Wood-Sparrow, who sings as if he were delighted at being left almost alone to warble and complain to the benevolent deities of the grove. He who in his youth has made frequent visits to these pleasant and solitary places, among the thousands of beautiful and sweet-scented flowers that spring up among the various spicy and fruit-bearing shrubs that unite to form a genuine whortleberry-pasture, — he only knows the unspeakable delights which are awakened by the sweet, simple notes of this little warbler.

The Wood-Sparrow is somewhat smaller than a Canary, with a pale chestnut-colored crown, above of a brownish hue, and dusky-white beneath. Though he does not seem to be a shy bird, I have never seen him in our gardens. The inmates of solitary cottages alone are privileged to hear his notes from their windows. He loves the plains and the hillsides which are half covered with a primitive growth of young pines, junipers, cornels, and whortleberry-bushes, and lives upon the seeds of grasses and wild lettuce, with occasional repasts of insects and fruits. His notes are mellow and plaintive, and, though often prolonged to a considerable length, seldom consist of more than one strain. He begins slowly and emphatically, as if repeating the syllable *de, de, de, de,* any number of times, increasing in rapidity, and at the same time sliding upward, by almost imperceptible gradations, about one or two tones on the musical scale.

WOOD-SPARROW'S SONG.

In the latter part of June, when this bird is most musical, he occasionally varies his song, by uttering a few chirps after the first strain, like the Canary, then recommencing it, and repeating it thus perhaps three or four times. I once heard a Canary that repeated this reiterated song of the Wood-Sparrow, and it seemed to me to surpass any notes I had ever heard before from this sweet little domesticated songster.

THE GROUND-ROBIN OR CHEWINK.

While listening to the notes of the Wood-Sparrow, we are constantly saluted by the agreeable, though less musical, notes of the Ground-Robin, an amusing little bird that confines himself chiefly to the edges of woods. This bird is elegantly spotted with white, red, and black, the female being of a bright bay color where the male is red. Every rambler knows him, not only by his plumage and his peculiar note, but also by his singular habit of lurking among the bushes, appearing and disappearing like a squirrel, and watching all our movements. It is with difficulty that a gunner can obtain a good aim at him, so rapidly does he change his position among the leaves and branches. In these motions he resembles the Wren. When he perceives that we are observing him he pauses in his song, and utters that peculiar note of complaint from which he has derived the name Chewink. The sound is more like *chewee*, accenting the second syllable.

The Chewink is a very constant singer during four months of the year, from the first of May. He is untiring in his lays, seldom resting for any considerable time from morn to night, being never weary in rain or in sunshine, or at noonday in the hottest weather of the season. His song consists of two long notes, the first about a third above the second, and the last part made up of several rapidly uttered liquid notes, about one tone below the first note.

SONG OF THE CHEWINK.

There is an expression of great cheerfulness in these notes, though they are not delivered with much enthusiasm. But music, like poetry, must be somewhat plaintive in its character to take strong hold of the feelings. I have never known any person to be affected by these notes as many are by those of the Wood-Sparrow. While employed in singing, the Chewink is usually perched on the lower branch of a tree, near the edge of a wood, or on the summit of a tall bush. He is a true forest bird, and builds his nest upon the ground in the thickets that conceal the boundaries of the wood.

The note of the Chewink and his general appearance and habits are well adapted to render him conspicuous, and to cause him to be known and remembered, while the Wood-Sparrow and the Veery might remain unobserved. Our birds are like our "men of genius." As in the literary world there is a description of mental qualities which, though of a high order, must be pointed out by an observing few before the multitude can appreciate them, so the sweetest songsters of the wood are unknown to the mass of the community, while many ordinary performers, whose talents are conspicuous, are universally known and admired.

### THE REDSTART AND SPECKLED CREEPER.

As we advance into the wood, if it be midday, or before the decline of the sun, the notes of two small birds will be sure to attract our attention. The notes of the two are very similar and as slender and fine as the chirp of a grasshopper, being distinguished from it only by a different and more pleasing modulation. These birds are the Redstart and the Speckled Creeper. The first is the more rarely seen. It is a bird of the deep forest, and shuns observation by hiding itself in some of the obscure parts of the wood. Samuels, however, has known a nest of the Redstart to be built and the young reared in a garden, and other authors consider the bird more familiar than shy. In general markings, that is, as we view the bird without particular examination, the Redstart is like the Chewink, though not more than half its size. It lives entirely on insects, darting out upon them from its perch like a flycatcher, and searching the foliage for them like a sylvian. Its song is similar to that of the Summer Yellow-Bird, so common in our gardens among the fruit-trees, but more shrill and feeble. The Creeper's note does not differ from it more than the notes of different individuals of the same species.

The Speckled Creeper takes its name from its habit of creeping like a Woodpecker round the branches of trees, feeding upon the insects and larvæ that are lodged in the crevices of the bark. It often leaves the wood and diligently manœuvres among the trees in our gardens and enclosures. The constant activity of the birds of this species affords proof of the myriads of insects that must be destroyed by them in the course of one season, and which, if not kept in check by these and other small birds, would, by their multiplication, render the earth uninhabitable by man.

### THE OVEN-BIRD.

While listening to the slender notes of these little sylvians, hardly audible amidst the din of grasshoppers, the rustling of leaves, and the sighing of winds among the tall oaken boughs, suddenly the space resounds with a loud, shrill song, like the sharpest notes of the Canary. The little warbler that startles us with this vociferous note is the Golden-crowned Thrush or Oven-Bird. This bird is confined almost exclusively to the woods, and is particularly partial to noonday, when he sings. There is no melody in his lay. He begins rather moderately, increasing in loudness as he proceeds, until his note seems to fill the whole wood. He might be supposed to utter the words *I see, I see, I see, I see*, emphasizing the first word, and repeating the two five or six times, growing louder and louder with each repetition. There is not a bird in the wood that equals this little piper in the energy with which he delivers his brief communication. His notes are associated with summer noondays in the deep woods, and when bursting upon the ear in the silence of noon, they disperse all melancholy thoughts as if by enchantment.

Samuels says ho has listonod to tho song of this bird at all hours of the night, in the mating and incubating season. The bird seems to soar into the air, and to sing while hovering in a slow descent. He has noticed the same habit in the Maryland Yellow-Throat. Dr. Brewer says the Oven-Bird "has two very distinct songs, each in its way remarkable." I have noticed that many species of birds are addicted occasionally to a kind of soliloquizing; warbling in a low tone, not very audibly and apparently for their own amusement. It is seldom that these soliloquizing notes bear any resemblance to the usual song of the bird; and I have heard them from the Chickadee and other birds that have no song.

The oven-shaped nest of this bird has always been an object of curiosity. It is placed upon the ground under a knoll of moss, or a tuft of weeds and bushes, and is neatly woven of long grass and fibrous roots. It is covered with a roof of the same materials, and a round opening is made at the side for entrance. The nest is so ingeniously covered with grass and assimilated to the surface around it, that it is not easily discovered. But it is said that the Cowbird is able to find it, and uses it as a depository for its eggs.

### THE GREEN WARBLER.

Those who are accustomed to rambling in the forest may have observed that pine woods are remarkable for certain collections of mosses which have cushioned a projecting rock or the decayed stump of a tree. When weary with heat and exercise, it is delightful to sit down upon one of these green velveted couches and take note of the objects immediately around us. We are then prepared to hear the least sound that pervades our retreat. Some of the sweetest notes ever uttered in the wood are distinctly heard only at such times; for when we are passing over the rustling leaves, the noise made by our progress interferes with the perfect recognition of all delicate sounds. It was when thus reclining, after half a day's search for flowers, under the grateful shade of a pine-tree, now watching the white clouds that sent a brighter daybeam into those dark recesses as they passed luminously overhead; then noting the peculiar mapping of the ground underneath the wood, diversified with mosses in swelling knolls, little islets of fern, and parterres of ginsengs and Solomon's-seals, I was first greeted by the pensive note of the Green Warbler, as he seemed to utter in supplicating tones, very slowly modulated, *Hear me, St. Theresa!* This strain, as I have observed many times since, is at certain hours repeated constantly for ten minutes at a time; and it is one of those melodious sounds that seem to belong exclusively to solitude.

Though these notes of the Green Warbler may be familiar to all who are accustomed to strolling in the wood, the bird is known to but few persons. Some birds of this species are constant residents during summer in the woods of Eastern Massachusetts, but the greater number retire farther north in the breeding season. Nuttall remarks of the Green Warbler: "His simple, rather drawling, and somewhat plaintive song, uttered at short intervals, resembles the syllables *te, de, deritsca*, pronounced pretty loud and slow, the tones proceeding from high to low. In the intervals, he was particularly busied in catching small cynips and other kinds of flies, keeping up a smart snapping of his bill, almost similar to the noise made by knocking pebbles together."

There is a plaintive expression in this musical supplication that is apparent to all who hear it, no less than if the bird were truly offering prayers to some tutelary deity. It is difficult to determine why a certain combination of sounds should affect one with an emotion of sadness, while another, under the same circumstances, produces a feeling of joy. This is a part of the philosophy of music which has not been explained.

### SONG OF THE GREEN WARBLER.

Hear me, St. The - re - sa.

### THE MARYLAND YELLOW-THROAT.

As we leave the forest and emerge into the open pasture, we hear a greater number of birds than in the darkness of the wood. More sounds are awake of every description, not only those of a busy neighboring population, but of domestic birds and quadrupeds. On the outside of the wood, if the ground be half covered with wild shrubs, you will hear often repeated the lively song of the Maryland Yellow-Throat. Like the Summer Yellow-Bird, he is frequently seen among the willows; but he is less familiar, and seldom visits the garden or pleasure-ground. The angler is startled by his notes on the rushy borders of a pond, and the botanist listens to them while peeping into some woodland hollow or bushy ravine. Even the woodcutter is delighted with his song, when, sitting upon a new-fallen tree, he hears the little bird from a near cornel-bush, saying, *I see, I see you, I see, I see you, I see, I see you.* These notes are not unlike those of the Brigadier, and are both lively and agreeable.

In its plumage the Yellow-Throat is very attractive. It is of a bright olive-color above, with a yellow throat and breast, and a black band extending from the nostrils over the eye. The black band and the yellow throat, are the marks by which the bird is readily identified. From its habits of perching low, frequenting the undergrowth near the edge of the wood, building upon the ground, and seldom visiting the higher branches of trees, it has obtained the name of Ground Warbler.

### THE SCARLET TANAGER.

When I was about seven years of age I first saw the Scarlet Tanager, lying dead in a heap of birds which had been shot by two Spaniards, who were my father's private pupils. The fine plumage of this bird soon attracted my attention. But it was long before I could feel reconciled to this slaughter, though delighted with the opportunity of examining the different birds in the heap. Since that time I have often found the Scarlet Tanager in the gamebags of young sportsmen; but I have seldom seen in the woods more than two or three birds of this species in any one season.

Low grounds and oaken woods are the Tanager's favorite habitats. It nestles in the deep forest, and builds a loosely constructed nest of soft grass and slender brush, forming a shallow basket which is lodged upon some horizontal bough of oak or pine. This bird, however, displays no skill as a basket-maker, hardly surpassing even the Turtle-Dove as an architect. The eggs are speckled on a ground of dull pea-green. The male Tanager sings with considerable power a sort of interrupted song, modulated a little after the manner of the Thrush. Samuels kept one confined six months in a cage, and in a week after its capture it submitted quietly to its confinement, and became tuneful. He compares its song to that of the Robin, mixed with some ventriloquial notes. We hear this bird in the deep wood more frequently than outside of it.

## THE FLICKER.

We are all familiar with the notes of this Woodpecker, that resemble the call-notes of the common Robin, but they are louder and more prolonged. Audubon compares them to the sounds of laughter when heard at a distance. According to the same writer the males woo the females very much after the manner of our common Doves. They build in holes in trees, but you never see them climbing a tree like other Woodpeckers. They take their food chiefly from the ground, and devour great quantities of ants.

The Flicker, though not attractive when seen at a distance, is found to have very beautiful plumage on examination. On the back and wings it is chiefly of a light brown, with black bands on the wing-feathers, giving them a kind of speckled appearance; a scarlet crescent on the back of the head, and a similar shaped black patch on the throat. The under surface of the wings is of a golden yellow. Hence it is sometimes called the Golden-winged Woodpecker. Samuels relates that if the eggs, which are of a pure white, be removed from the nest while the bird is laying, she will continue to lay like a common hen. He has known this experiment to be tried until the bird had laid eighteen or twenty eggs, though her usual number is but six.

## THE ROSE-BREASTED GROSBEAK.

We must pass out of the woods again, where we can bask in the sunshine, and obtain a view of fields and farms, to hear the voice of the Rose-breasted Grosbeak. This bird was not an acquaintance of my early years. Certain changes of climate or soil, either here or in its former habitats, have caused it to be a regular sojourner in New England for twenty years past, and the species arrive every year in increased numbers. Formerly their residence was chiefly confined to the Middle States. Now we may see them frequently every summer, but not in familiar places or in those which are very solitary. I have seen them many times in Medford woods, and in those near Fresh Pond in Cambridge, and in Essex County.

The first time I heard the note of the Grosbeak I mistook it for the song of the Golden Robin, prolonged, varied, and improved in an unusual degree. I soon, however, discovered the bird, and thought his lively manners, no less than his brilliant notes, were like those of the Golden Robin. His song is greatly superior to that of the Redbird or Cardinal Grosbeak, which is only a repetition of two or three sweet notes, like *che-hoo, che-hoo, che-hoo*, rapidly delivered, the last note of each two about a third lower than the first. In the South he is joined by the Mocking-Bird, which all day tiresomely repeats these notes of the Cardinal.

The Rose-breasted Grosbeak is classed among our nocturnal songsters by those who are familiar with its habits. Samuels has heard it frequently in the night, and says of its song that it is "a sweet warble with various emphatic passages, and sometimes a plaintive strain exceedingly tender and affecting." This description seems to me very beautiful and accurate. Mr. S. P. Fowler thinks this bird is not heard so frequently by night as by day, though it often sings in the light of the moon. The moon, indeed, seems to be the source of inspiration to all nocturnal songsters. Though I once mistook the song of this Grosbeak for that of the Golden Robin, lately I have thought it more like the native song of the Mocking-Bird, and not inferior to it in any respect. He utters but few plaintive notes. They are mostly cheerful, melodious, and exhilarating. They are modulated somewhat like those of the Purple Finch, delivered more loudly and with a great deal more precision.

# FLOWERS AS EMBLEMS.

THE custom of emblemizing flowers, which has prevailed among all nations, springs from a native passion of the human mind. To the fancy they are persons, objects of friendship and love, having the semblance of our virtues and affections. If we speak of them with passionate regard, it is because we thus personify them and clothe them with human and even divine qualities. The virtues we admire in the characters of our friends we are delighted to behold symbolized in flowers. Hence those representing modesty, humility, delicacy, and purity are our favorites, while we seldom long admire the gaudy and showy flowers. We prize them in proportion as they are suggestive of some pleasing moral sentiment. Hence a white flower, which is without beauty of color, often gains more of our admiration than another similar one of beautiful tints.

Wordsworth habitually views the minor works of nature through this moral coloring, and loves to speak the praises of the common and simple garden flowers. Like a true poet, he sees in them more to awaken pleasant and salutary thoughts than in those which are prized at floral exhibitions. He has woven many delightful emblematic images with flowers, and through them has conveyed important sentiments of a moral and religious kind. He considers the daisy, which is scattered widely in England over every field and near every footpath, and which is also cultivated at cottage-windows in many different countries, as a " pilgrim of nature," whose home is everywhere. He thinks there abides with this little plant some concord with humanity; and that those who are easily depressed may learn a lesson from it. It will teach them by its cheerful example how to find a shelter in every climate and under all conditions of adversity, engaging the affections of all no less by its modest beauty than by its capacity of living and thriving, and remaining bright and cheerful under all circumstances of culture or neglect.

He also praises, in another poem, the small celandine. He greets it as the prophet of spring and its attractions; and speaks of the thrifty cottager who stirs seldom out of doors, and who is charmed with the sight of this humble flower by reason of its happy augury of the year. He commends it for its kindly and unassuming disposition. Careless of its neighborhood, we see its pleasant face in wood and meadow, in the rustic lane and in the stately avenue, on the princely domain and in the meanest place upon the highway. It is pleased and contented in all situations, and the poet glows in his description of its unpretending virtues. He rebukes the gaudy flowers that will be seen whether we would see them or not, and considers them as exemplifying the pride of worldlings; and again he extols the virtues of the small celandine.

In another poem he compares the ambitious, who, without more than ordinary talents or merit, aspire to some lofty station, to a tuft of fern on the summit of a high rock. It is a miserable thing, "dry, withered, light, and yellow," that endeavors to soar with the tempest and expose itself to observation; but all its importance belongs to its position. We wonder how it came there, and how it is able to keep its place, while plants of superior qualities would be unable to transport themselves thither; and if by accident they should arrive at such a height, they could not sustain it. The fern by its meanness accomplishes what, if it possessed a nobler nature, would be impossible. Thus, he continues, mean men, never doubting their own merit or capacity, and unscrupulous of the means they use to elevate themselves or to keep their place, rise to eminences which men of genius and integrity could not attain, because they scorn the actions that would insure them success.

The rose, in all ages, has been regarded as the emblem of beauty and virtue, having in addition to its visual attractions a fragrance that always endures. The Hebrew and classical writers have associated this flower with certain divine qualities which are held up for our love and reverence. The lily is no less celebrated, being frequently mentioned in Holy Writ to adorn a parable or to improve the force of some poetic image. Among all nations it is a chosen symbol of meekness and modesty, and it is more frequently celebrated in lyric poetry than any other flower, because it is the semblance, in the highest degree, of those qualities which are favorite themes of the poets. Its paleness is typical of delicacy, while its drooping habit renders it a true emblem of sorrow. It is the metaphorical image of the meek and passive virtues, while the perfume it sends abroad may be compared to the influence of a good man's life.

I have said nothing of the language of flowers, which seems in general to have only a slight foundation in nature. It is rather the result of an agreement to use certain flowers to signify certain words or ideas arbitrarily applied to them. It is indeed but an agreeable form of writing by cipher. In some cases this language is founded on a legend or a poetic fable, in others on the emblematic characters of the flowers. Thus, the violet signifies modesty, because its colors are soft, and the flower seems to hide itself from observation. In like manner the sensitive plant is expressive of purity, because it shrinks from the touch; and the balsam of impatience, because its capsules snap in the hand that is put forth to gather them. Let us not deride the harmless amusements that spring from this philological use of flowers, nor despise the ingenuity that invented them. A bouquet that conveys an affectionate message from a young lover to his mistress possesses a charm in her sight which genius could hardly express in the finest verses.

Flowers serve a more needful purpose in the economy of nature than we are prone to imagine; and they produce more effect on the dullest minds than many even of the most susceptible would acknowledge. But it is not an uncommon habit, especially among the ignorant, to ridicule the study of flowers and those who are devoted to it. On the other hand, they do not despise the occupation of the florist, because it brings him money. Others consider botany a trifling pursuit, worthy the attention only of persons of effeminate habits; but I have never been able to learn that these objectors are contemners of any of those fashionable habits which are confessedly enervating and destructive of mental and physical power. Nothing can enervate that actively employs the mind and exercises the body at the same time, as may be said of the outdoor study of botany and that of any other branch of natural history. They are the most invigorating of all intellectual pursuits. Nor is the study of flowers the less worthy of attention, though we admit that it exercises the imagination and fancy more than it stores the mind with knowledge. The same charge may be brought against the pursuit of any of the fine arts.

The botanist, however, does not study flowers merely as beautiful objects. As a scientific observer, he finds in them the exponents of the laws of vegetation, which can be understood only by the keenest perceptions. Hence the fact that among botanists may be named some of the greatest men who have lived. As a moral and poetic observer, he discerns in flowers, not mere gems sparkling on the bosom of Nature, but so many living beings, looking up to him from the greensward, and down upon him from the trees and cliffs, and inspiring him with a feeling of sympathy with all the visible world. What can be more worthy of study than this beautiful assemblage of living things, whose relations to each other and to men and animals unfold a thousand singular mysteries, whose forms and colors produce the most delicate conceptions of art, and whose metaphorical characters have rendered them the very poetry of nature! Religion and virtue, science, painting, and poetry, all have their readings in these brilliant pets of the florist and toys of children. The stars of heaven do not convey to our minds a more vivid conception of the mysteries of the universe than the flowers that sparkle in the same countless numbers on the earth.

Let us imagine that the earth had been created without flowers; that the greensward was sprinkled with no violets in the opening of the year, and that May flung around her footsteps neither daisies nor cowslips; that summer called out no blossoms upon the trees, and that autumn bound with his ripened sheaves neither asters nor goldenrods, and looked through his frosty eyelids upon neither gentians nor euphrasia! Let us imagine that the dews cherished nothing fairer than the green foliage of herbs and trees, and that the light of morning, which now unfolds the splendor of millions of tinted corols, sparkled only in the crystal dewdrops; that the butterfly looked in vain for its counterpart among the plants that now offer it their allurements, and that the bee was not one of the living forms of nature, because the fields produced no flowers for its sustenance! Who would not feel that some unknown blessing was denied us? Who would not believe that there was some imperfection in the order of nature?

What fanciful image of happiness is not associated with flowers, — the delight of infant ramblers in the sunshine of May; the reward of their searchings in the meadows among brambles and ferns; infantile honors and decorations for the brows of childhood; the types of their budding affections and the materials for their cheerful devices; the ornaments of young May-queens and the joy of their attendants; the fair objects of their quest in the sunny borders of fragrant woods; the pride of their simple ambition when woven into garlands of love! How blank would the earth be to childhood without flowers! How destitute the fields of beauty and nature of poetry!

But Nature, who set light in heaven to beam with every imaginable hue, has not made us sensitive to beauty without bestowing upon the earth those forms which, like the letters of a book, convey to the mind an infinity of delightful thoughts and conceptions. Hence flowers are made to spring up in wood and dell, by solitary streams, in moss-grown recesses, near every path that glides through the meadow, and in every green lane that wanders through the forest; and Nature has given them an endless variety of forms, colors, and deportment, that by their different expressions they may awaken every agreeable passion of the soul. There is no place where their light is not to be seen. The inhabitant of the South beholds them in trees looking down upon him like the birds; the man of the North sees them embossed in verdure, under the protection of trees and rocks. Insects sip from their honey-cups the nectar of their subsistence, during a life as ephemeral as that of the blossom they plunder; and the summer gales rejoice in their sweets, with which they have laden their wings. Morning greets them when she wakes, and sees them spread out their petals to the light of the sun, all glowing with beauty when the dews that sleep nightly in their bosoms steal silently back to heaven; and every day is relieved of its weariness by the myriads that brighten when it approaches, and sweeten with their fragrance the transitory visits of each fleeting hour.

Where is the mind so impassive that it is not animated by the presence of flowers and made hopeful by their gayety? Where is the eye that does not see them, and note their comeliness, and wish that they might never droop or decay? Where is the lover that does not view them as partaking of his own passion, and looking fair for the sake of her for whom they seem to be created? The young bride, when garlanded with their wreaths, feels that the virtues that should reside in her heart have shed their grace upon her through these fair symbols; and mourners, when they see them clustering round the tomb of a departed friend, worship them as lights of heaven, foreshowing in their sleep and resuscitation the soul's immortality!

# PICTURESQUE ANIMALS.

It may be observed that in pictures, when a certain effect is required, an animal is often introduced whose character and habits correspond with the scenery, or the sentiment to be awakened. A scene in nature without some such accompaniment often fails in producing any emotion in the mind. A heron standing on the borders of a solitary mere, a kingfisher sitting on the leafless branch of a tree that extends over the tide, a woodpecker climbing the denuded branch of an oak, yield to the respective scenes in which they are represented a life and a character which could not be so well expressed without them. A few cows grazing on a grassy slope, a dog reposing at the doorstep of a cottage, or a cat quietly slumbering inside of the window, are each suggestive of pleasant images of rural life, and add greatly to the interest of the scene. The majority of animals require to be viewed in connection with certain other objects to acquire a picturesque expression; but there are others which are endowed with this quality in a remarkable degree, and need only to be seen in any situation to awaken a certain agreeable train of images.

Among birds the owl is often represented in engravings, when it is designed to impart to the scene a character of desolation. We often see this bird accompanying a picture of ruins or of a deserted house, and in poetry he is introduced to awaken certain peculiar trains of thought. Thus the poet Gray, when he would add a desolate expression to his description of evening, speaks of the owl as complaining to the moon of such as molest his ancient solitary reign. The allusion to his nocturnal habits and to his solitary dominions brings still more vividly to mind those qualities with which the image of the bird is associated. His appropriate habitations aro the ruined tower, the ancient belfry, or the hollow of an old tree. In all such places the figure of the owl is deeply suggestive of those fancies which are awakened by the sight of ancient dilapidated buildings, crumbling walls, and old houses supposed to be the residence of wicked spirits which are permitted to visit the earth.

It is on account of these dreary and poetic associations that the owl is so truly picturesque. He is often seen, in paintings and engravings, perched on an old gateway, or on one of the bars of an old fence, whose posts, leaning obliquely, show that they have been heaved by the frosts of many winters. In certain situations our slumbers are sometimes disturbed by the peculiar hooting of this bird, that awakens in the mind the gloomy horrors of midnight. His nocturnal and solitary habits, the unearthly tones and modulation of his voice, his practice of frequenting rude and desolate places and haunted houses, have caused his image to be intimately connected with mystery and gloomy forebodings of evil. The very stillness of his flight yields a sort of mysterious character to the bird; all these circumstances, combined with his fabled reputation for wisdom, and his demure and solemn expression of countenance, have conspired to render the owl one of the most picturesque of all living creatures.

The bat is another creature, in some respects, of similar habits and reputation. Like the owl, it naturally seeks, for its retreat during the day, those unfrequented places where it is not liable to be disturbed, and has acquired a character and expression in harmony with the scenes it frequents. But it is remarkable that while the owl has obtained an emblematical character for wisdom, the bat is regarded as the emblem of guilt. He is represented as shunning the broad eye of day, and as flying out on leathern wing, when the dusky shades of evening may serve to hide him from detection. The sight of the bat, however, is far from awakening in our minds the idea of guilt; but his image is strongly suggestive of the pleasant serenity of evening, as the butterfly reminds us of summer fields and flowers. Our ideas of the bat are somewhat grotesque; and when, after the graceful swallow has retired to rest, we observe his irregular and zigzag flight, we are unavoidably reminded of his peculiar hideous formation, from which the idea of making him an emblem of guilt probably originated. It would seem as if he hid himself during the day, lest his relationship to a race of beings now almost banished from the earth might be discovered. His emblematical character does not prevent his forming an interesting feature in a rural scene. Hence in pictorial representations of evening we see the last rays of the sun streaming upward in beautiful radiations from behind a hill, while the bat is flitting about an old house in a rude and rather quiet landscape.

All animals are picturesque which are consecrated to poetry. In English descriptive poetry the lark is as familiar to us as the rose that clambers around the cottage door. The unrivalled brilliancy of his song, which, by description, is impressed on our minds with a vividness almost like that of memory, and its continuance after he has soared to an immense height in the air, cause him to be allied in our minds with the sublimity of heaven and with the beauty and splendor of morning. I never had an opportunity to witness the flight of the skylark; but I have always imagined that the sentiment of sublimity must greatly enhance the pleasure with which we gaze upon his flight and listen to his notes.

The very minuteness of an object soaring to such a sublime elevation gives us an idea of some almost supernatural power, and his delightful song would seem to be derived from heaven, whither he takes his flight while giving utterance to it. We have no skylarks in America; but our common snipes, during the month of May, are addicted to this habit of soaring, as I have remarked in another essay, for a few hours after sunset. I have often watched them in former times, and when witnessing their spiral flight upwards to a great elevation, and listening to their distinct but monotonous warbling after they have arrived at the summit of their ascent, I have been conscious of an emotion of sublimity from a spectacle which might be supposed too trivial to produce any such effect. The picturesque character of the lark is apparent only when he is represented in his soaring flight. There is nothing peculiar in the appearance of this bird as in that of the owl. The sight of him aloft in the heavens is necessary, therefore, to suggest the idea of his habits and to make his true character apparent to the mind.

Among the animals mentioned by certain writers as possessing in an eminent degree those qualities which appertain to the picturesque, is the ass. This point in his character is attributed very erroneously to his shaggy and uncouth appearance. It may assist in heightening the expression of the animal; but there are various romantic and poetical ideas associated with his figure, to which this quality is mainly attributable. If it were owing to his rude and rough exterior, the baboon and the hyena would be as picturesque as the ass. No such ideas, however, are associated with these animals. The ass derives much of this character from his connection with the incidents of romance and history. He is the beast of burden most frequently mentioned in the Old Testament, in the Fables of Æsop, and in the writings of Oriental travellers. As Dugald Stewart has observed, we associate him with the old patriarchs in their journeys to new lands; and we have often seen him forming an important figure in old paintings and engravings. It is not his shaggy coat and uncouth appearance that yield him his picturesque character, so much as the interesting scenes and adventures with which his figure is associated.

The same remarks may be applied with equal propriety to the goat. He is the animal of mountain scenery, and the sight of him brings to mind a variety of romantic incidents connected with such landscape. He is often represented as standing on precipitous heights and browsing upon dangerous declivities. He is, in fact, one of the dumb heroes of dangerous adventure. With the inhabitants of mountainous countries, as among the Alps and the Highlands of Scotland, the goat is the domesticated animal that supplies them with milk. The hardiness and activity of the goat, his frequent introduction into pictures of Alpine scenery, and his habit of finding sustenance in wild regions and fastnesses where no other animal could live, combine to render his image strongly suggestive of rusticity and the simple habits of mountaineers.

It is common to regard the uncouthness of the appearance of these animals as the quality from which they derive their picturesque expression. It is much more probable that, on account of the absence of beauty of color, smoothness, and symmetry, the imagination is left more entirely to the influence of the poetic and traditional images connected with these animals. In this way it may be explained why rudeness is, to a certain extent, a negative picturesque quality, because it leaves the imagination entirely to the suggestions of the scene; whereas, if it were very beautiful, the mind would be more agreeably occupied in surveying its intrinsic beauties than in dwelling upon its more poetical relations to certain other ideas and objects.

Why is the horse not a picturesque animal, it may be asked, but on account of the sleekness of his appearance? I am persuaded that his sleekness stands in the way of this expression, only so far as it causes him to be associated with fashion and the pomp and pride of wealth. Hence, it must be allowed that the only horses that have this quality are shaggy ponies and cart-horses. This proves only that their rough exterior is the indication of the rusticity of their habits, not that it is an intrinsic quality of the picturesque, which has indeed no intrinsic properties, like beauty, but is founded on association. Were the case reversed, and were animals to become sleek when engaged in rustic employments, and rough and hairy when fed and combed and pampered by wealthy and lordly masters, in that case the smoothest animals would be the most picturesque. The squirrel, which is a sleek and graceful animal, is, in spite of these qualities, more picturesque than the rough and rusty looking rat. In this instance the principle is reversed, because the smoothness and gracefulness of the squirrel are associated with his interesting habits of playfulness and agility, while running about from branch to branch among his native groves. On the contrary, the smooth and symmetrical horse cannot, by any pictorial accompaniments, be made so expressive as the rough and homely ass.

I have just alluded to the squirrel as one of the most picturesque of the smaller animals; but it is worthy of notice that it must be represented in its native habitats to possess this character in full force. Though a squirrel in a cage is a beautiful object, especially when turning his revolving grate by the rapid motions of his feet, yet a picture of one in that situation would have none of that suggestiveness of poetical and agreeable fancies that renders a scene picturesque. In a representation of a little cottage in the woods nothing could add more to its pleasing pastoral expression than the figure of a squirrel running along on a stone-wall or on the branch of an old tree. The sight awakens all those poetical images which are associated with life in the fields. Place the squirrel in a cage, and it reminds us only of the town, and expresses nothing that is agreeable to a poetic fancy. Every wild animal must appear to be enjoying its freedom, or the representation of it would fail in giving delight. The same is true of the human race. While persons of the laboring classes add to the interesting character of a scene in nature, a single figure, male or female, in fashionable apparel, destroys the whole effect. Hence, almost all the representations of picnics fail in awakening any poetic emotions.

A shepherd, when properly represented with his crook, which is his staff of office, and surrounded by the animals of his charge, his faithful dog, the rustic cottage, the sheepfold, and the general rude scenery of nature, is always picturesque. But his appearance must be entirely that of a shepherd, without any of the ways or the gear of a man of the town. I have seen a picture of two young shepherds in the mountains, in which the characteristic qualities of the scene are entirely destroyed by a certain genteel or finical air and expression observed in their countenance and attitudes. Instead of rustic shepherds we see two young men, each with a crook, sitting and reclining upon a rock. They are very neatly dressed, and look as if they were young sprigs of the nobility, who had gone into the mountains for a few days, merely to play shepherd; so nicely is their hair arranged, that the longitudinal parting is distinctly seen, caused by the smoothing away of the hair on each side of the head. The expression of their faces corresponds with the rest of their appearance; one, in particular, having that look of conscious self-satisfaction which we often observe in a silly fop of the town. The very manner in which he leans his head upon his thumb and fingers betrays his concern lest he should spoil the arrangement of his hair. How strange that the painter of this piece should not have seen that all these little trifles completely ruined the picturesque character of his painting!

One of the most interesting engravings I have seen represents a peasant-girl, in the neat and simple attire of her own humble station in life, in the act of bearing a pitcher of water which she has just dipped from a rustic well. How easily might the designer have ruined the whole expression of this piece, either by making the well an elegant and fanciful structure or by making the damsel a fine lady in her silks and laces. The sight of a picnic party assembled together in the woods and pastures is always pleasing; but, as I have already intimated, it fails in interest when represented on canvas, because, with all the fine images connected with it, it savors of the vanity of fashionable or rather of town life. After witnessing one of these scenes, while journeying leisurely in a chaise on a pleasant day in October, I chanced to see a group of little country girls, in the simplest apparel, gathering nuts under a tree. What a crowd of pleasant recollections of the past was immediately awakened by the sight! "There," exclaimed my companion, "is a scene for a painter. Such a little group in a picture would afford us inexpressible delight. Yet, were I to join either party, I should prefer to be one of the other company at the picnic." "For the very plain reason," I replied, "that in the latter company you would expect to find some intelligent persons who would be interesting companions. But this is not what we look for in a picture, which pleases in proportion to the simplicity of its characters."

These remarks might be indefinitely extended; but each new example would serve only to repeat the illustration of the same principle. In no other engravings do we see the picturesque more clearly exemplified than in the vignettes which are found in books published early in the last century. Since luxury has extended into the circle of the middle and industrious classes, the simplicity of their habits has been destroyed, and artists, when drawing their designs from the manners of these classes, have failed in producing pictures equal in poetic expression to those which were made one hundred years ago. It is apparent, for example, that the ancient straw beehive, surrounded by its swarm, formerly introduced into vignettes as emblematical of industry, is decidedly picturesque; while the modern patent structures, constructed for purposes of economy, would, in fanciful engravings, excite ideas no more poetical than we should find in a modern revolving churn. Modern customs and improvements are rapidly sweeping away from the face of the earth everything that is poetic or picturesque. It may be urged, however, that the sum of human happiness has been proportionally increased. This I am inclined to doubt; and to maintain, on the contrary, that just in proportion as we depart from the simple habits of the early era of civilization, do we create wants that cannot be gratified, and lose those tastes which are most promotive of happiness and in harmony with the designs of nature and of providence.

# JUNE.

ALREADY do we feel the influence of a more genial sky; a maturer verdure gleams from every part of the landscape, and a prouder assemblage of wild-flowers reminds us of the arrival of summer. The balmy southwest reigns the undisturbed monarch of the weather; the chill breezes rest quietly upon the serene bosom of the deep; and the ocean, as tranquil as the blue canopy of heaven, yields itself to the warm influence of the summer sun, as if it were conscious of the blessing of his beams. The sun rides, like a proud conqueror, over three quarters of the heavens, and, as if delighted with his victory over the darkness, smiles with unwonted complacency upon the beautiful things which are rejoicing in his presence. Twilight refuses to leave the brows of night, and her morning and evening rays meet and blend together at midnight beneath the polar sphere. She twines her celestial rosy wreaths around the bosoms of the clouds, that rival in beauty the terrestrial garlands of summer. The earth and the sky seem to emulate each other in their attempts to beautify the temples of nature and of the Deity; and while the one hangs out her drapery of silver and vermilion over the sapphirine arches of the firmament, the other spangles the green plains and mountains with living gems of every hue, and crowns the whole landscape with lilies and roses.

The mornings and evenings have acquired a delightful temperature, that invites us to rise prematurely from our repose, to enjoy the greater luxury of the balmy breezes.

The dews hang heavily upon the herbage, and the white frosts have gone away to join the procession of the chill autumnal nights. The little modest spring flowers are half hidden beneath the prouder foliage of the flowers of summer; the violets can hardly look upon us from under the broad leaves of the fern; and the anemones, like some little unpretending beauty in the midst of a glittering crowd, are scarcely observed as they are fast fading beneath the shade of the tall shrubbery. The voice of the early song-sparrow and the tender warbling of the blue-bird are but faintly audible amidst the chorus of louder musicians; the myriads of piping creatures are silent in the wet places, and the tree-frogs, having taken up their song, make a constant melodious croaking, after nightfall, from the wooded swamps. The summer birds have all arrived; their warbling resounds from every nook and dell; thousands of their nests are concealed in every grove and orchard, among the branches of the trees, or on the ground beneath a tuft of shrubbery; egg-shells, of various hues, are cast out of their nests, and the callow young lie in the open air, exposed to the tender mercies of the genial month of June.

The season of anticipation has passed away; the early month of fruition has come; the hopes of our vernal morning have ripened into realities; we no longer look into the future for our enjoyments, but we revel at length in all those pleasures from which we expected to derive a perfect satisfaction. The month of June is emblematical of the period of life that immediately succeeds the departure of youth, when all our sources of enjoyment are most abundant, and our capacity for higher pleasure has attained maturity, and when the only circumstance that damps our feelings is the absence of that lightness of heart arising from a hopeful looking forward to the future. Our manhood and our summer have arrived, but our youth and our spring have gone by; and though we have the enjoyment of all we anticipated, yet with the fruition hope begins to languish, for in the present exists the fulness of our joys. The flowery treasures, foretokened by the first blue violet, are blooming around us; the melodious concert, to which the little song-sparrow warbled a sweet prelude in March, is now swelling from a full band of songsters, and the sweet summer climate that was harbored by an occasional south-wind has arrived. But there is sadness in fruition. With all these voluptuous gales and woodland minstrelsies, we cannot help wishing for a renewal of those feelings with which we greeted the first early flower and listened to the song of the earliest returning bird.

Nature has thus nearly equalized our happiness in every season. When our actual joys are least abundant, fancy is near at hand, to supply us with the visions of those pleasures of which we cannot enjoy the substance; filling our souls in spring with the hope of the future, comforting us in autumn with the memory of the past, and amusing us in winter with a tranquil retrospection of the whole year and the pleasant watching for the dawn of another spring.

A total change has taken place in the aspect of the woods since the middle of the last month. The light, yellowish green of the willows and thorns, the purple of the sumach, and the various hues of other sprouting foliage have ripened into a dark uniform verdure. The grass, as it waves in the meadows, gleams like the billows of the ocean; and the glossy surfaces of the ripe leaves of the trees, as they tremble in the wind, glitter like millions of imperfect mirrors in the light of the sun. The petals of the fading blossoms are flying in all directions, as they are scattered by the fluttering gales, and cover, like flakes of snow, the surface of the orchards.

The flowers of innumerable forest-trees are in a state of maturity, and the yellow dust from their flower-cups, scattered widely over the earth, may be seen after showers, covering the edges of the beds of dried water-pools, in yellow circular streaks.

The pines and other coniferous trees are in flower during this month; and the golden hues of their blossoms contrast beautifully with the deep verdure of their foliage. These trees, like others, shed their leaves in autumn; but it is the foliage of the preceding year that falls, leaving that of the last summer still upon the trees. This foliage is very slowly perishable, and covers the earth where it falls, during all the year, with that brown, smooth, and fragrant carpet, which is characteristic of a pine wood. Among the flowers which are conspicuous on this brown matted foliage is the purple lady's-slipper, whose inflated blossoms often burst upon the sight of the rambler, as if they had risen up by enchantment. In similar haunts the trientalis, unrivalled in the peculiar delicacy of its flowers, that issue from a single whorl of pointed leaves, supported upon a tall and slender footstalk, never fails to attract the attention of the botanist and the lover of nature.

Our gardens, during the first of this month, exhibit few exotics more beautiful than the Canadian rhodora, an indigenous shrub, which is at this time in full flower in the wild pastures. It is from two to five feet in height, and its brilliant purple flowers, unrivalled in delicacy, appear on the extremities of the branches, when the leaves are just beginning to unfold. It is rendered singularly attractive by the contrast between its purple hues, of peculiar resplendency, and the whiteness of the flowers of almost all other shrubs, at this season. This plant, by its flowering, marks the commencement of summer, and may be considered an apt symbol of the brilliant month of June.

## JUNE. 151

June is the month of the arethusas, — those charming flowers of the peat-meadows, — belonging to a tribe that is too delicate for cultivation. Like the beautiful birds of the forest, they were created for Nature's own temples; and the divinities of the wood, under whose invisible protection they thrive, will not permit them to join with the multitude that grace the parterre. The cymbidium, of a similar habit, the queen of the meadows, with larger flowers and more numerous clusters; the crimson orchis, that springs up by the river-sides, among the myrtle-like foliage of the cranberry and the nodding panicles of the quaking-grass, like a spire of living flame; and the still more rare and delicate white orchis, that, hidden in deep mossy dells in the woods, seldom feels the direct light of the sun, — are all alike consecrated to solitude and to Nature, as if they were designed to cheer the hearts of her humble votaries with the sight of a thing of beauty that has not been appropriated for the exclusive adornment of the garden and the palace.

The rambler may already perceive a difference in the characters of the flowers of this month and of the last. In May the prominent colors were white and the lighter shades of purple and lilac, in which the latter were but faintly blended. In June the purple shades predominate in the flowers, except those of the shrubs, which are mostly white. The scarlet hues are seldom seen until after midsummer. The yellows seem to be confined to no particular season, being conspicuous in the dandelion, ranunculus, and coltsfoot of spring; in the potentilla, the senecio, and the loosestrife of summer; and in the sunflower, golden-rod, and many other tribes of autumn. Blue is slightly sprinkled through all the seasons.

One of the most charming appearances of the present month, to one who is accustomed to the minute observation of Nature's works, is the flowering of the grasses.

Though this extensive tribe of plants is remarkable in no instances for the brilliancy of its flowers, yet there are few that exhibit more beauty in their aggregations; some rearing their flowers in a compact head, like the herd's-grass and the foxtail; others spreading them out in an erect panicle, like a tree, as the orchard-grass and the common redtop; others appearing with a bristling head, like wheat and barley; and a countless variety of species, with nodding panicles, like the oat and the quaking-grass. The greater number of the gramineous plants are in flower at the present time, and there are no other species, save the flowerless plants, which afford more attractions to those who examine nature with the discriminating eye of science.

He who is accustomed to rambling is now keenly sensible of that community of property in nature, of which he cannot be deprived. The air of heaven belongs equally to all, and cannot be monopolized; but the land is apportioned into tracts belonging to different owners, and the many perhaps do not own a rood. Yet to a certain extent, and in a very important sense, the earth, the trees, the flowers, and the landscape are common property. He who owns a fine garden possesses but little advantage over him who is without one. We are all free in this country to roam over the wide fields and pastures; we can eat of the fruits of the earth, and feast our eyes on the beauties of nature, as well as the owner of the largest domain. A man is not poor who, while he obtains the comforts of life, is thus capable of enjoying the blessings of nature. His property is not circumscribed by fences and boundary lines. All the earth is his garden, — cultivated without expense and enjoyed without anxiety. He partakes of these bounties which cannot be confined to a legal possessor, and which Providence, as a compensation to those who are worn with toil or harassed with care, spreads out to gladden them with renewed hopes and to warm their hearts with gratitude and benevolence.

June is, of all months of the year, the most delightful period of woodland minstrelsy. With the early birds that still continue their warbling, the summer birds have joined their louder and more melodious strains. Early in the morning, when the purple light of dawn first awakens us from sleep, and while the red rays that fringe the eastern arches of the sky with a beautiful tremulous motion are fast brightening into a more dazzling radiance, we hear from the feathered tribe the commencement of their general hymn of gladness. There is first an occasional twittering, then a single performance from some early waker, then a gradual joining of new voices, until at length there is a full chorus of song. Every few minutes some new voice joins in the concert, as if aroused by the beginners and excited by emulation, until thousands of melodious voices seem to be calling us out from sleep to the enjoyment of life and liberty.

After the sun has risen nearly to meridian height, the greater number of the birds that helped to swell the anthem of morn discontinue their songs, and a comparative silence prevails during the heat of the day. The vireo, however, warbles incessantly, at all hours of daylight, from the lofty tree-tops in the heart of the villages; the oriole is still piping at intervals among the blossoms of the fruit-trees; and the merry bobolink never tires during the heat of the day, while singing and chattering, as in ecstasy, above and around the sitting-place of his wedded mate. At the hour of the sun's decline the birds renew their songs; but the more familiar species that linger about our orchards and gardens are far less musical at sunset than at sunrise. I suppose they may be annoyed by the presence of men, who are more accustomed to be out at a late hour in the evening than at an early hour in the morning.

The hour preceding dusk in the evening, however, is the time when the thrushes, the most musical of birds, are loudest in their song. Several different species of this tribe of musicians, at a late hour, are almost the sole performers. The catbird, with a strain somewhat similar to that of the robin, less melodious, but more varied and quaint in its expression, is then warbling in those places where the orchards and the wildwood meet and are blended together. The red-thrush, a bird still more retired in its habits, takes his station upon a tree that stands apart from the wood, and there pours forth his loud and varied song, which may be heard above every other note. A little deeper in the woods, near the borders of streams, the veeries, the last to become silent, may be heard responding to one another, with their trilled and exquisite notes, unsurpassed in melody and expression, from the sun's early decline until the purple of twilight has nearly departed. During all this time and the greater part of the day, in the solemn depths of the forest, where almost all other singing-birds are strangers, resounds the distinct, peculiar, and almost unearthly warbling of the hermit-thrush, who recites his different strains with such long pauses and with such a varied modulation that they might be mistaken for the notes of several different birds.

At nightfall, though the air is no longer resonant with song, our ears are greeted with a variety of pleasing and romantic sounds. In the still darkness, apart from the village hum, may be heard the frequent fluttering of the wings of night birds, when the general silence permits their musical vibrations to resound distinctly from different distances, during their short, mysterious flights. These sounds, to which I used to listen with ravishment in my early days, are more suggestive than music, and always come to my remembrance, as one of the delightful things connected with a summer evening in the country. At the same time, in my rambles after sunset, I have often paused to hear the responsive chirping of the snipes, in the open plains, during their season of courtship; and to watch their occasional whirling flight, as with whistling wings they soar like the lark into the skies, to meet and warble together, above the darkness that envelops the earth. With the same whirling flight, they soon descend to the ground, and commence anew their responsive chirping. These alternate visits to the earth and the sky are continued for several hours. There is nothing very musical in the chirping of these birds; and their warbling in the heavens, when they have reached the summit of their ascent, is only a somewhat monotonous succession of sounds. But when, at this later time of life, I chance to hear a repetition of their notes, the whole bright page of youthful adventure is placed vividly before my mind. It is only at such times that we feel the full influence of certain sounds of nature in hallowing the period of manhood with a recollection of early pleasures and a renewal of those feelings that come upon the soul like a fresh breeze and the sound of gurgling waters to the weary and thirsty traveller.

The evenings are now so delightful that it seems like imprisonment to remain within doors. Odors, sights, and sounds are at present so grateful and tranquillizing in their effects upon the mind, and so suggestive of all the bright period of youth, that they cannot be regarded as the mere pleasures of sense. The sweet emanations from beds of ripening strawberries, from plats of pinks and violets, from groves of flowering linden-trees, full of myriads of humming insects, from meadows odoriferous with clover and sweet-scented grasses, all wafted in succession with every little shifting of the wind, breathe upon us an endless variety of fragrance. Then the perfect velvety softness of the evening air; the various melodies that come from every nook, tree, rock, dell, and fountain; the notes of birds, the chirping of insects, the hum of bees, the rustling of aspen leaves, the bubbling of fountains, the dashing of waves and waterfalls, and the many beautiful things that greet our vision from earth, sea, and sky, — all unite, as it were, to yield to mortals who hope for immortality a foretaste of the unspeakable joys of paradise.

# PLEA FOR THE BIRDS.

In the beginning, according to the testimony of the "Wisdom of Solomon," all things were ordered in measure, number, and weight. The universe was balanced according to a law of harmony no less wise than beautiful. There was no deficiency in one part or superfluity in another. As time was divided into seasons and days and years, the material world was arranged in such a manner that there should be a mutual dependence of one kingdom upon another. Nothing was created without a purpose, and all living things were supplied with such instincts and appetites as would lead them to assist in the great work of progression. The kingdoms of nature must ever remain thus perfectly adjusted, except for the interference of man. He alone, of all living creatures, has power to turn the operations of nature out of their proper course. He alone is able to transform her hills into fortifications and to degrade her rivers to commercial servitude. Yet, while he is thus employed in revolutionizing the surface of the earth, he might still work in harmony with nature's designs, and end in making it more beautiful and more bountiful than in its pristine condition.

In the wilderness we find a certain adjustment of the various tribes of plants, birds, insects, and quadrupeds, differing widely from that which prevails over a large extent of cultivated territory. In the latter, new tribes of plants are introduced by art, and nature, working in harmony with man, introduces corresponding tribes of insects, birds, and quadrupeds. Man may with impunity make a change of the vegetable productions, if he but allows a certain freedom to Nature in her efforts to supply the balance which he has disturbed. While man is employed in restocking the earth with trees and vegetables, Nature endeavors to preserve her harmony by a new supply of birds and insects. A superabundance of either might be fatal to certain tribes of plants. I believe the insect races to be as needful in the order of creation as any other part of Nature's works. The same may be said of that innumerable host of plants denominated weeds. But while man is endeavoring to keep down superfluities, he may, by working blindly, cause the very evil he designs to prevent. It is not easy to check the multiplication of weeds and insects. These, in spite of all direct efforts to check them, will increase beyond their just mean. This calamity would not happen if we took pains to preserve the feathered tribes, which are the natural checks to the multiplication of insects and weeds. Birds are easily destroyed: some species, indeed, are already nearly exterminated; and all are kept down to such a limit as to bear no just proportion to the quantity of insects that supply them with food.

Although birds are great favorites with man, there are no animals, if we except the vermin that infest our dwellings, that suffer such unremitted persecution. They are everywhere destroyed, either for the table or for the pleasure of the chase. As soon as a boy can shoulder a gun, he goes out, day after day, in his warfare of extermination against the feathered race. He spares the birds at no season and in no situation. While thus employed, he is encouraged by older persons, as if he were ridding the earth of a pest. Thus do men promote the destruction of one of the blessed gifts of Nature.

If there be proof that any race of animals was created for the particular benefit of mankind, this may certainly be said of birds. Men in general are not apt to consider how greatly the sum of human happiness is increased by certain circumstances of which they take but little note. There are not many who are in the habit of going out of their way or pausing often from their labors to hear the song of a bird or to examine the beauty of a flower. Yet the most indifferent would soon experience a painful emotion of solitude, were the feathered race to be suddenly annihilated, or were vegetation to be deprived of everything but its leaves and fruit. Though we may be accustomed to regard these things as insignificant trifles, we are all agreeably affected by them. Let him who thinks he despises a bird or a flower be suddenly cast ashore upon some desert island, and after a lonely residence there for a season, let one of our familiar birds greet him with a few of its old accustomed notes, or a little flower peep out upon him with the same look which has often greeted him by the wayside in his own country, and how gladly would he confess their influence upon his mind!

But there is a great deal of affectation of indifference toward these objects that is not real. Children are delighted with birds and flowers; women, who have in general more culture than men, are no less delighted with them. It is a common weakness of men who are ambitious to seem above everything that pleases women and children to affect to despise the singing of a bird and the beauty of a flower. But even those who affect this indifference are not wholly deaf or blind. They are merely ignorant of the influence upon their own minds of some of the chief sources of our pleasures.

It is not entirely on account of their song, their beauty, and their interesting habits, that we set so high a value upon the feathered tribes. They are important in the general economy of Nature, without which the operation of her laws would be disturbed, and the parts in the general harmony would be incomplete. As the annihilation of a planet would produce disturbance in the motions of the spheres, and throw the celestial worlds out of their balance, so would the destruction of any species of birds create confusion among terrestrial things. Birds are the chief and almost the only instruments employed by Nature for checking the multiplication of insects which otherwise would spread devastation over the whole earth. They are always busy in their great work, emigrating from place to place, as the changes of the seasons cut off their supplies in one country and raise them up in another. Some, like the swallow tribe, seize them on the wing, sailing along the air with the velocity of the winds, and preserving it from any excess of the minute species of atmospheric insects. Others, like the creepers and woodpeckers, penetrate into the wood and bark of trees, and dislodge the larvæ before they emerge into the open air. Beside these birds that do their work by day, there are others, like the whippoorwill tribe, that keep their watch by night, and check the multiplication of moths, beetles, and other nocturnal insects.

Man alone, as I have before remarked, can seriously disturb the operations of Nature. It is he who turns the rivers from their courses, and makes the little gurgling streams tributary to the sluggish canal. He destroys the forests, and exterminates the birds after depriving them of their homes. But the insects, whose extreme minuteness renders them unassailable by his weapons, he cannot destroy, and Nature allows them to multiply and become a scourge to him, as if in just retribution for his cruelty to the feathered races who are his benefactors.

In the native wilderness, where man has not interfered

with the harmonious operations of Nature, the insects are kept down to a point at which their numbers are not sufficient to commit any perceptible ravages. The birds, their natural destroyers, are allowed to live, and their numbers keep pace with the insects they devour. In cultivated tracts, on the contrary, a different state of things exists. Man has destroyed the forests, and raised up gardens and orchards in their place. The wild pasture has become arable meadow, and the whortleberry grounds have been changed into cornfields. New races of beetles and other insects, which are attached to the cultivated vegetables, increase and multiply in the same proportion. If man would permit, the birds that feed upon these insects would keep pace with their increase, and prevent the damage they cause to vegetation. But, too avaricious to allow the birds to live, lest they should plunder fruit enough to pay them the wages for their useful labors, he destroys the exterminator of vermin, and thus, to save a little of his fruit from the birds, he sacrifices his orchards to the insects.

If any species of birds were exterminated, those tribes of insects which are their natural food would become exceedingly abundant. Inasmuch as the atmosphere, if the swallows were to become extinct, would be rendered unfit for respiration, by an increased multitude of gnats and smaller insects; so, were the sparrow tribes to become extinct, vegetation would immediately suffer from an increase of caterpillars, curculios, and other pests of our orchards. We may say the same of other insects with relation to other birds. It is therefore plainly for the interest of the farmer and the horticulturist to use all means for the preservation of birds of every species. There is no danger likely to arise from their excessive multiplication. The number of each species cannot exceed that limit beyond which they could not be supplied with their proper and natural food. Up to this limit, if they were preserved, our crops would be effectually secured from the ravages of insects. The country would probably support double the present number of every species of birds, which are kept down below their proper limits by accident, by the gun of the sportsman, and by the mischievous cruelty of boys.

Most of the smaller kinds of birds have a disposition to congregate around our villages. We seldom find a robin or a sparrow, during breeding-time, in the deep forest. The same may be said of the insects that serve them for food. There are certain tribes that chiefly frequent the wild woods; these are the prey of woodpeckers and their kindred species. There are others which are abundant chiefly in our orchards and gardens; these are the prey of bluebirds, sparrows, wrens, and other common and familiar birds.

Man has the power to diminish the multitudes of insects that desolate the forests and destroy his harvests; but this can be effected only by preserving the birds, and Nature has endowed them with an instinct that leads them to congregate about his habitations, as if she designed them to protect him from the scourge of noxious vermin, and to charm his ears by the melody of their songs. Hence every tract which is inhabited by man is furnished with its native singing-bird, and man is endowed with a sensibility that renders the harmony of sounds necessary to his happiness. The warbling of birds is intimately associated with everything that is beautiful in nature. It is allied with the dawn of morning, the sultry quiet of noon, and the pleasant hush of evening. There is not a cottage in the wilderness whose inmates do not look upon the birds as the chief instruments of Nature to inspire them with contentment in their solitude. Without their merry voices, the silence of the groves, unbroken save by the moaning of the winds, would be oppressive; the fields would lose half their cheerfulness, and the forest would seem the very abode of melancholy. Then let our arms, designed only for self-defence, no longer spread destruction over the plains; let the sound of musketry no longer blend its discord with the voices of the birds, that they may gather about our habitations with confidence, and find in man, for whose pleasure they sing and for whose benefit they toil, a friend and a protector.

# BIRDS OF THE PASTURE AND FOREST.

## II.

### THE HERMIT-THRUSH.

THE bird whose song I describe in this essay has always seemed to me to be the smallest of the Thrushes. But as I have never killed any bird for the purpose of learning its specific characters, I am liable to be mistaken in many points of identification. It has been my habit from my earliest years, whenever I heard a note that was new and striking, to watch day after day, until I discovered the songster, and, having always had excellent sight, I have never used a telescope. The bird whose notes I describe below, when I have seen it upon a tree or upon the ground, has seemed to conform more nearly to the description given in books of the Hermit-Thrush, both in size and color, than to that given of the Wood-Thrush.

The notes of this bird are not startling or readily distinguished. Some dull ears might not hear them, unless their attention was directed to the sounds. They are loud, liquid, and sonorous, and they fail to attract attention only on account of the long pauses between the different strains. We must link all these strains together to enjoy the full pleasure they are capable of affording, though any single one alone would entitle the bird to considerable reputation as a songster. He also sings as much at broad noonday as at any other time, differing in this respect from the Veery, who prefers the twilight of morn and even. In another important respect he differs

UNIV. OF
CALIFORNIA

from the Veery, which is seldom heard except in swamps, while the Hermit almost invariably occupies high and dry woods.

The Hermit-Thrush delights in a shady retreat; he is indeed a true anchorite; he is evidently inspired by solitude, and sings no less in gloomy weather than in sunshine. Yet I think he is no lover of twilight, though pleased with the darkness of shady woods; for at the time when the Veery is most musical, he is generally silent. He is remarkable, also, for prolonging his musical season to near the end of summer. Late in August, when other birds have become silent, he is almost the only songster in the wood.

The song of the Hermit consists of several different strains, or bars, as they would be called in the gamut. I have not determined the exact number, but I am confident there are seven or eight, many of them remarkable for the clearness of their intonations. After each strain he makes a full pause, perhaps not more than three or four seconds, and the listener must be very attentive, or he will lose many of the notes. I think the effect of this sylvan music is somewhat diminished by the pauses or rests. It may be said, however, that during each pause our susceptibility is increased, and we are thus prepared to be more deeply affected by the next notes. Some of these are full and sonorous, like the sound of a fife; others lisping, and somewhat like the chink made by shaking a few thin metallic plates in your hand. This lisping strain always comes regularly in its course. I can imagine that if all these different strains were warbled continuously, they would not be equalled by the song of any bird with which I am acquainted.

Some parts of Nuttall's description of the song of the Hermit, if it be identical with the species called by him the Song-Thrush, are incorrect. It is not true that his different strains or those of the Wood-Thrush "finally blend together in impressive and soothing harmony, becoming more mellow and sweet at every repetition." Any one strain never follows another, without a full pause between them. I think Nuttall has described the song of the Veery, mistaking it for a part of that of the Song-Thrush. One of the enunciations which he attributes to the Song-Thrush is equally remarkable and correct. I allude to "the sound of *ai-ro-ee*, peculiarly liquid, and followed by a trill." The song invariably begins with a clear fife sound, as *too, too, tillere illere*, rising from the first about three musical tones to the second, and making the third and fourth words rather sharp and shrill. We seldom, however, hear more than one low note in a strain, as *too, tillere illere;* afterwards, beginning with the low note *too*, follows the sound of *ai-ro-ee*, like the notes of the common chord. The fourth bar is a lisping strain resembling the sounds made by shaking thin metallic plates in the hand; the fifth, a trilling like the notes of the Veery, — *tillillil, tillillil, tillillil.* There are several other bars consisting of a slight variation of some one of those I have described. I have not been able to determine the order in which the several strains succeed one another. I feel confident, however, that the bird never repeats any one strain, save after two or three others have intervened.

The Wood-Thrush is a larger bird than the Hermit, more common in our woods, having a similar song, containing fewer strains, delivered with less precision and moderation, and with shorter intervals between the high and the low notes. In their general habits the two species differ very slightly.

## THE VEERY, OR WILSON'S THRUSH.

The Veery is perceptibly larger than the Hermit, and is marked in a similar manner, save that the back has more of an olive tinge. He arrives early in May, and is first heard to sing during some part of the second week of that month. He is not one of our familiar birds; and unless we live in close proximity to a wood that is haunted by a stream, we seldom hear his voice from our doors and windows. He sings neither in the orchard nor the garden. He shuns the town, and reserves his wild notes for those who live in cottages by the woodside. All who have once become familiar with his song await his arrival with impatience, and take note of his silence in midsummer with regret. Though his song has not the compass and variety of that of the Hermit, it is more continuous and delivered with more fervor. Until this little bird arrives, I feel as an audience do at a concert before the chief singer appears, while the other performers are vainly endeavoring to soothe them by their inferior attempts.

The Veery is more shy than any other important singing-bird except the Hermit. His haunts are solitary woods, usually in the vicinity of a pond or a stream. Here, especially after sunset, he warbles his few brilliant but plaintive strains with a peculiar cadence, and fills the whole forest with music. It seems as if the echoes were delighted with his notes, and took pleasure in passing them round with multiplied reverberations. I am confident that this little warbler refrains from singing when others are vocal, from the pleasure he feels in listening either to his own notes or to the melodious responses which others of his own kindred repeat in different parts of the wood. Hence, he chooses the dusk of evening for his tuneful hour, when the little chirping birds are silent, that their voices may not interrupt his chant.

At this hour, during a period of nine or ten weeks, he charms the evening with his strains, and often prolongs them in still weather until after dusk, and whispers them sweetly into the ear of Night.

His song, though loud for so small a bird, is modulated with such a sweet and flowing cadence that it comes to the ear like a strain from some elfin source. It seems at first to be wanting in variety. I formerly thought so, while at the same time I was puzzled to account for its enchanting effect on the mind of the listener. The same remark may be applied to the human voice. I suppose I am not the only person who can remember certain female voices, which, with limited compass and execution, do, by a peculiar native modulation, combined with great simplicity, affect the listener with emotions such as no *prima donna* could produce. Having never heard the Nightingale, I can draw no comparison between that bird and the Veery. But neither the Mocking-Bird, nor any other bird in our woods, utters a single strain to be compared in sweetness and expression to the five bars of the simple song of the Veery.

Were we to attempt to perform these notes upon a musical instrument, we should fail from the difficulty of imitating their peculiar trilling and the liquid ventriloquial sounds at the end of each strain. The whole is warbled in such a manner as to produce on the ear the effect of harmony, and to combine in a remarkable degree the two different qualities of brilliancy and plaintiveness. The former effect is produced by the first notes of each strain, which are sudden and on a high key; the second by the graceful chromatic slide to the termination; which is inimitable and exceedingly solemn. I have sometimes imagined that a part of the delightful influence of these notes might be ascribed to the cloistered recesses in which they are delivered. But I have occasionally heard them while the bird was singing from a tree near the heart of a village, when they were equally delightful and impressive.

In my early days, when I was at school, I lived near a grove that was vocal with these Thrushes. It was there I learned to love their song more than any other sound in nature, and above the finest strains of artificial music. Since then I have seldom failed to make frequent visits to their habitats, to listen to their notes, which cause full half the pleasure I derive from a summer evening ramble.

Dr. Brewer does not so highly estimate the song of the Veery, but Mr. Ridgway differs from him. "To his ear," says Dr. Brewer, "there was a solemn harmony and a beautiful expression which combined to make the song of this bird surpass that of all the other American Wood-Thrushes." I have found the nests of this species very near the ground, also upon a mound of grass and sticks, and on a bush. Their eggs are of a greenish-blue.

THE CATBIRD.

Fond of solitude, but not averse to the proximity of human dwellings, if the primitiveness of some of the adjacent wood remains; avoiding the deep forest and the open pastures, and selecting for his habitat the edge of a wooded swamp, or a fragment of forest near the low grounds of a cultivated field, the Catbird may be seen whisking among the thickets, often uttering his complaining mew, like the cry of a kitten. Still, though attached to these wet and retired situations, he is often very familiar, and is not silenced by our presence, like the Veery. His nest of dry sticks is sometimes woven into a currant-bush in a garden that adjoins a swamp, and his quaint notes may be heard, as if totally unmindful of the nearness of his human foe. The Catbird is not an inveterate singer. He seldom makes music his sole employment; though at any hour of the day, from dawn till evening twilight, he may occasionally be heard singing and complaining.

Though I have been all my life familiar with the notes and manners of the Catbird, I have not been able to discover that in his native woods he is a mocker. He seems to me to have a definite song, unlike that of any other songster, except the Red-Thrush. It is not made up from the notes of other birds, but is as unique and original as the song of the Robin or the Linnet. In the song of any bird we may detect occasional strains that resemble those of some other species; but the Catbird gives no more of these imitations than we might reasonably regard as accidental. The truth is, that the Thrushes, though delightful songsters, have inferior powers of execution, and cannot equal the Finches in learning and performing the notes of other birds. Even the Mocking-Bird, compared with many other species, is a very imperfect imitator of any notes which are rapid and difficult of execution. He cannot give the song of the Canary; yet I have heard a caged Bobolink do this to perfection.

The modulation of the Catbird's song is somewhat similar to that of the Red-Thrush, and I have found it sometimes difficult to determine, from the first few notes, whether I was listening to the one or the other; but after a moment I detected one of those quaint utterances that distinguish the notes of the Catbird. I am confident that no man would mistake this song for that of any other species except the Red-Thrush; and in this case his mistake would soon be corrected by longer listening. The Red-Thrush has a louder and fuller intonation, more notes that resemble speech, or that may be likened to it, and some fine guttural tones which the other never utters.

I repeat that I have not any proof, from my own observation, that the Catbird is a mocker. Dr. Brewer says, on the other hand, that it is a very good imitator of simple notes and strains. He has heard it give excellent imitations of the whistling of the Quail, the clucking of a Hen, the notes of the Pewee, and those of the Ground-Robin, repeating them with such exactness as to deceive the birds that were imitated. He has known the Catbird call off a brood of young chickens, to the great annoyance of the old hen.

The Catbird is said to be very amusing when confined in a cage. A former neighbor of mine, who has reared many birds of this species in a cage, informed me that when tamed they sing better than in their native woods. He taught them not only to imitate the notes of other birds, but to sing tunes. This is an important fact; but we must confess that the wild birds and the wild-flowers are more interesting in their native haunts than in aviaries or conservatories. Though I have no sensibility that would prevent my depriving a bird of its freedom by placing it in a comfortable prison, where it would suffer neither in mind nor body, I should not keep one in a cage for my own amusement, caring but little to watch its ways except in a state of freedom.

The mewing note of the Catbird, from which his name was derived, has been the occasion of many misfortunes to his species, causing them to share that contempt which is so generally felt towards the feline race; and that contempt has been followed by persecution. The Catbird has always been proscribed by the New England farmers, who from the first settlement of the country have entertained a prejudice against the most useful of our birds, which are also the most mischievous. Even the Robin has been frequently in danger of proscription. The horticulturists, who seem to consider their cherries and strawberries and favorite insipid pears of more importance than the whole agricultural crop of the States, have made several efforts to obtain an edict of outlawry against him. These repeated onslaughts have induced the friends of the Robin to examine his claims to protection, and the result of their investigations is demonstrative proof that it is one of the most useful birds in existence. The Catbird and all the Thrushes are similar to the Robin in their habits of feeding, but are not sufficiently numerous to equal it in the extent of their services.

### THE RED-THRUSH.

After we have grown tired of threading our way through the half-inundated wood-paths in a swamp of red-maple and northern cypress, where there is twilight at broad noonday, and where the only sounds we hear are the occasional sweet notes of the Veery, now and then a few quaint utterances from the Catbird, and the cawing of Crows, high up in the cedars, we emerge into the upland under the bright beams of noonday. The region into which we enter is an open pasture of hill and dale, more than half covered with wild shrubbery, and displaying an occasional clump of trees. There, perched upon the middle branch of some tall tree, the Red-Thrush, the rhapsodist of the woods, may be heard pouring forth his loud and varied song, often continuing it without cessation for half an hour. His notes do not, like those of the Finches and many other birds, have a beginning, a middle, a turn, and a close, as if they were singing the words of a measured hymn. The notes of the Red-Thrush are more like a voluntary for the organ, in which, though there is a frequent repetition of certain strains, the close of the performance comes not after a measured number of notes.

The Red-Thrush has many habits similar to those of

the Catbird, but he is not partial to low grounds. He prefers the dry hill and upland, and those places which are half cleared, and seems averse to deep woods. Still, though less of a hermit than the Catbird, he is also less familiar. He dislikes the proximity of dwelling-houses, and courts the solitude of open fields and dry hills distant from the town. This bird probably owes its shyness and timidity to the desperation with which the species have been hunted by men who are unwilling that the birds shall take any pay for the services they perform; and who, to save a dozen cherries from a bird, would sacrifice the tree to mischievous insects. Modern civilized society bears the besom of a devastation greater than the world has yet seen, and when it has completed its work, and destroyed every bird and animal that is capable of doing any service to agriculture, man will perish too, and the whole earth become a combined Sahara and wilderness of Mount Auburns.

The Red-Thrush builds in a low bush, or more frequently upon the ground under a bush. I think he sings at some distance from his nest, selecting for his musical moments the branch of a tree that projects over a rustic roadside. As the roadside supplies a greater abundance of larvæ than the wild pastures, it may be that after having taken his repast, he perches near the place where he obtained it. He is not partial to any certain hour for singing, but is most musical in fine and bright weather. I can always hear him where he dwells in the vocal season, morning, noon, and evening. When employed in song, he makes it his exclusive occupation, and sings, though moderately, with uninterrupted fervor. In this respect he is distinguished above almost all other species. I have observed, however, that if he be disturbed while singing, he immediately becomes silent and may not renew his song under an hour.

The Red-Thrush is considered by many persons the finest songster in the New England forest. Nuttall says "he is inferior only to the Mocking-Bird in musical talent." I doubt his inferiority except as a mocker. He is superior to the Mocking-Bird in variety, and is surpassed by him only in the sweeter intonations of some of his notes. But no person grows tired of listening to the Red-Thrush, who constantly varies his notes, while the Mocking-Bird tires us with his repetitions, which are often continued to a ludicrous extreme. Perhaps I might give the palm to the Mocking-Bird, were it not for his detestable habit of imitation. But when this habit is considered, I do, without hesitation, place the Red-Thrush above him as a songster, and above every other bird with whose notes I am acquainted. If I were listening to a melodramatic performance, in which all were perfect singers and actors, I should prefer the *prima donna* to the clown, even if the clown occasionally gave a good imitation of her voice.

When we are in a thoughtful mood, the song of the Veery surpasses all others in tranquillizing the mind and yielding something like enchantment to our thoughts. At other times, when strolling in a whortleberry pasture, it seems to me that nothing can exceed the simple melody of the Wood-Sparrow. But without claiming for the Red-Thrush, in any remarkable degree, the plaintiveness that distinguishes these pensive warblers, his song in the open field has a charm for all ears, and can be appreciated by the dullest of minds. Without singing badly he pleases the millions. He is vocal at all hours of the day, and when thus employed, devotes himself entirely to song with evident enthusiasm.

It would be difficult, either by word or by musical notation, to give to one who has not heard the song of the Red-Thrush a correct idea of it. This bird is not a rapid singer. His performance is a sort of *recitative*, often resembling spoken words rather than musical notes, many of which are short and guttural. He seldom whistles clearly, like the Robin, but he produces a charming variety of tone and modulation. Some of his notes are delivered rapidly, but every strain is followed by a momentary pause, resembling the discourse of a man who speaks fast, but hesitates after every few words. He is rapid, but not voluble.

An ingenious shoemaker, named Wallace, whom I knew in my early days, and who, like many others of his craft when they worked alone or in small companies in their own shops, and not by platoons as in a steam factory, was a close observer of nature and mankind, gave me the following words as those repeated by the Red-Thrush: "Look up, look up, — Glory to God, glory to God, — Hallelujah, Amen, Videlicet."

Thoreau, in one of his quaint descriptions, gives an offhand sketch of the bird, which I will quote: "Near at hand, upon the topmost spray of a birch, sings the Brown-Thrasher, or Red Mavis, as some love to call him, — all the morning glad of your society (or rather I should say of your lands), that would find out another farmer's field if yours were not here. While you are planting the seed, he cries, 'Drop it, drop it, — cover it up, cover it up, — pull it up, pull it up, pull it up.'"

The Red-Thrush is most musical in the early part of the season, or in the month succeeding his arrival about the middle of May; the Veery is most vocal in June, and the Song-Thrush in July. The Catbird begins early and sings late, and fills out with his quaint notes the remainder of the singing season, after the others, save the Song-Thrush, have become silent.

# THE FLOWERLESS PLANTS.

As a tribe of vegetable curiosities, pleasantly associated with cool grots, damp shady woods, rocks rising in the midst of the forest, with the edges of fountains, the roofs of old houses, and the trunks and decayed branches of trees, may be named the flowerless plants. Few persons know the extent of their advantages in the economy of vegetation ; still less are they aware how greatly they contribute to the beauty of some of the most beautiful places in nature, affording tints for the delicate shading of many a native landscape, and an embossment for the display of some of the fairest flowers of the field. The violet and the anemone, that peep out upon us in the opening of spring, have a livelier glow and animation when embosomed in their green beds of moss; and the arethusa blushes more beautifully by the side of the stream when overshadowed by the broad pennons of the umbrageous fern. The old tree with its mosses wears a look of freshness in its decay, the bald rock loses its baldness with its crown of lichens and ferns, and every barren spot in the pasture or by the wayside is enlivened and variegated by the carpet of flowerless plants, that spread their green gloss and many-colored fringes over the surface of the soil.

Mosses enter into all our ideas of picturesque ruins ; for they alone are evidence that the ruins are the work of time. An artificial ruin can have no such accompaniment until time has hallowed it by veiling its surface with these memorials. They join with the ivy in adorning the relics of ancient grandeur, and spread over the perishable works of art the symbols of a beauty that endureth forever. While they are allied to ruins, and remind us of age and decay, they are themselves glowing in the freshness of youth, and cover the places they occupy with a perpetual verdure. They cluster around the decayed objects of nature and art, and are themselves the nurseries of many a little flower that depends on them for sustenance and protection. Though they bear no flowers upon their stems, they delight in cherishing in their soft velvet knolls the wood-anemone, the houstonia, the cypripedium, and the white orchis, — the nun of the meadows, — whose roots are imbedded among the fibres of the peat-mosses, and derive support from the moisture that is accumulated around them. Nature has provided them as a shield to many delicate plants, which, embowered in their capillary foliage, are enabled to sustain the heat of summer and the cold of winter, and remain secure from the browsing herds.

Winter, which is a time of sleep with the higher vegetable tribes, is a season of activity with many of the flowerless plants. There are certain species of mosses and lichens that vegetate under the snow, and but few of the mosses are at all injuriously affected by the action of frost. By this power of living and growing in winter, they are fitted to act as protectors to other plants from the vicissitudes of winter weather, and by their close texture they prevent the washing away of the soil from the declivities into the valleys. They answer the double purpose of catching the floating particles of dust and retaining them about their roots, and of preventing any waste from the places they occupy. Finding in them the same protection which is afforded by the snow, or by the matting of straw provided by the gardener, there are many plants that vegetate under their surface, secure from the alternate action of freezing and thawing in winter, and of drought in summer. Hence certain plants blossom more luxuriantly in a bed of mosses than in the unoccupied soil.

The mosses are seldom found in cultivated lands. As they grow entirely on the shallow surface, the labors of the tiller of the soil are fatal to them. They delight in old woods, in moist barren pastures, in solitary moorlands, and in all unfrequented places. In those situations they remain fresh and beautiful, while they prepare for the higher vegetable tribes many a barren spot, that must otherwise remain forever without its plant. They are, therefore, the pioneers of vegetable life; and Nature, when she selects an uncongenial tract to be made productive of fruits or flowers, covers the surface with a close texture of moss, and variegates it with lichens, before she strews the seeds of the higher plants to vegetate among their roots. The wise husbandman, who, by a careful rotation of crops, causes his land to be constantly productive, is but an humble imitator of Nature's great principle of action.

The mosses have never been made objects of extensive cultivation by our florists. Every rambler in the wild wood knows their value and their beauty, which seem to have been overlooked by the cultivator. They undoubtedly possess qualities that might be rendered valuable for purposes of artificial embellishment. There is no tree with foliage of so perfect a green tint as that of the moss which covers the roofs of very old buildings. The mossy knolls in damp woods are peculiarly attractive on account of their verdure, and the fine velvety softness of their pleasantly rounded surface. Though the mosses produce no flowers, the little germs that grow on the extremities of their hair-like stems are perfect jewels. With them, however, it is the stem that exhibits the most beauty of hues, varying from a deep yellow to a clear and

THE FLOWERLESS PLANTS. 179

lively claret or crimson, while the termination is green or brown. I have nothing to say of the physiology of their propagation. I treat of mosses only as they are beautiful objects of sight, and useful agents in unfolding and distributing the bounties of Nature. This tribe furnishes no sustenance to man or to any other animal. Those eatable plants which are called by the name of mosses are either lichens or sea-weeds. Nature, who, with a provident hand, renders many of her productions capable of supplying a manifold purpose in her economy, has limited the agency of the mosses to a few simple and beautiful services. They perform, under her invisible guidance, for the field and the forest, what is done by the painter and the embosser for the works of the builder of temples and palaces.

The ferns have fewer picturesque attractions than the mosses; but like the latter, they are allied with the primitive wilds of nature, with gloomy swamps, which they clothe with verdure, and with rocky precipices, on whose shelvy sides they are distributed like the tiles on the roof of a house. They resemble mosses in their dissimilarity to common vegetable forms; and their broad wing-like leaves or fronds are the conspicuous ornaments of wet woods and solitary pastures which are unvisited by the plough. By their singular appearance we are reminded of the primitive forms of vegetation on the earth's surface, and of the luxuriant productions of the tropics.

The ferns are for the most part a coarse tribe of plants, having more beauty in their forms than in their texture. In temperate latitudes it is only their leaf or frond that is conspicuous, their stems being either prostrate or subterranean. Yet in some of the species nothing can be more beautiful than the ramifications of their fronds. In their arrangements we may observe a perfect harmony and regularity, without the formality that marks the compound leaves of other plants. Herein Nature affords an example of a compound assemblage of parts, in a pleasing uniformity that far exceeds the most ingenious devices of art. Apparently similar arrangements are seen in the leaves of the poison hemlock, the milfoil, and the Roman wormwood; but their formality is not so beautifully blended with variety as that of the compound-leaved ferns.

In tropical countries some of the ferns are woody plants, attaining the size of trees, rising with a branchless trunk over fifty feet in height, and then spreading out their leaves like a palm-tree. Hence they are singularly attractive objects to the traveller from the North, by the sight of which he seems to be carried back to the early ages of the world, before the human race had a foothold upon the earth. Here we know them only as an inferior tribe in relation to size, the tallest seldom exceeding two or three feet in height. Everything in their appearance is singular, from the time when they first push up their purple and yellow scrolls above the surface of the soil, covered with a sort of downy plumage, to the time when their leaves are spread out like an eagle's wings, and their long spikes of russet flowers, if they may be so called, stand erect above the weeds and grasses, forming a beautiful contrast with the pure summer greenness of all other vegetation.

There are few plants that exceed in beauty and delicacy of structure the common maiden-hair. The main stem is of a glossy jet, and divided into two principal branches, that produce in their turn several other branches from their upper side, resembling a compound pinnate leaf without its formality. In woods in the western part of this State is a remarkable fern called the walking leaf. It derives its name from a singular habit of striking root at the extremities of the fronds, giving origin to new plants, and travelling along in this manner from one point to another. There is only one climbing fern among our native plants. Equally beautiful and rare, it is found only in a few localities all the way from Massachusetts to the West Indies. Unlike other ferns in its twining habit, it has also palmate leaves, with five lobes, and bears its fruit in a panicle, like the osmunda. But we need not search out the rare ferns for specimens of elegance or beauty. The common polypody, with its minutely divided leaves, covers the sides of steep wooded hills and rocky precipices, and adorns with a beautiful evergreen verdure their barren slopes, otherwise destitute of attractions. The ferns and the mosses are peculiarly the ornaments of waste and desert places, clothing with their verdure desert plains and rough declivities.

I have always attached a romantic interest to the seaweeds, whose forms remind me of the haunts of the Nereids, of the mysterious chambers of the ocean, and of all that is interesting among the deep inlets of the sea. Though flowerless, they are unsurpassed in the delicate arrangement of their branches, and the variety of colors they display. We see them only when broken off from the rocks on which they grew, and washed upon the shore, where they lie, after a storm, like flowers scattered upon the greensward by the scythe of the mower. When branching out in the perfection of their forms, underneath the clear briny tide, they are unsurpassed by few plants in elegance. The artist has taken advantage of their peculiar branching forms and their delicate hues, and weaves them into chaplets of many beautiful designs.

The sea-weeds seem to be allied to the lichens, and are considered by some botanists as the same plants modified by growing under water, and tinted by the iodine and bromine which they imbibe from the sea.

The lichens are the lowest tribe in the scale of vegetation. They make their appearance on naked rocks, and clothe them with a sort of fringe, holding fast on the rock for security, and deriving their chief sustenance from the atmosphere and particles of dust wafted on the winds and lodged at their roots. They have properly, however, no roots, neither have they leaves or stem; yet they are almost infinitely varied in their forms, hues, and ramifications. They grow in all places which are exposed to air and moisture, on the surface of rocks, old walls, fences, posts, and on branches of trees. Some of the species are foliaceous, resembling leaves without branches or any distinct or regular outline, and they are found mostly on rocks. Others are erect and ramified like trees and shrubs, but without anything that represents foliage. Such is that common gray lichen that covers our barren hills, which is a perfect hygrometer, crumbling under the feet in dry weather, and yielding to the step like velvet, whenever the air contains moisture. In similar places, and growing along with it, is found one of the *hepatic* mosses that produces those little tubercles — the fructification of the plant — resembling dots of sealing-wax, eagerly sought by artists who manufacture designs in moss. But the most beautiful lichens are those which are pendent from trees, consisting of branching threads, of an ash-green color, and bearing little circular shields at their extremities. These lichens give character to moist woods and low cedar-swamps, where they hang like funereal drapery from the boughs and deepen the gloom of their solitudes.

Lichens, though inhabiting all parts of the earth, are particularly luxuriant in cold climates, thriving in extreme polar latitudes, where not another plant can live. Nature seems to have designed them as an instrument for preparing every barren spot with the means of sustaining the more valuable plants. Not only do they cause a gradual accumulation of soil by their decay, but they actually feed upon the rocks by means of oxalic acid that exudes from their substance. By this process the surface of the solid rock is changed into a soil fitted for the nutrition of plants. After the lichens have perished, the mosses and ferns take root in the soil that is furnished by their decay. One vegetable tribe after another grows to perfection and perishes, but to give place to its more noble successor, until a sufficient quantity of soil is accumulated for the growth of a forest of trees. In such order may the whole earth have been gradually covered with plants, by the perishing of one tribe after another, leaving its substance for the support of a superior tribe, until the work of creation was completed.

Among the grotesque productions of nature, the *fungi*, or mushroom tribe, ought undoubtedly to be named as the most remarkable, attaining the whole of their growth in the space of a few days, and sometimes of a few hours. They are simple in their parts, like what may be supposed to have been the earliest productions of nature. They have no leaves or flowers or branches. They will grow and continue in health without light, requiring nothing but air and moisture above their roots. Though so low in the scale of vegetation, they are not without elegance of form and beauty of color, and are remembered in connection with dark pine woods, where, forming a sort of companionship with the monotropas, they are particularly luxuriant. Neither are they deficient in poetical interest, as these plants are the cause of those fairy rings that attract attention by their mysterious growth in circles on the greensward in the pastures.

The mushrooms vary extremely in their forms and sizes. Some are as slender as the finest mosses, tinted with gold and scarlet, and almost transparent. Others resemble a parasol, with their upper surface of a brilliant straw-color, dotted with purple, and their under surface of rose or lilac. They seem to riot in all sorts of beautiful and peculiar shapes and combinations. But the greater number are remarkable only for their grotesque forms, as if intended as a burlesque upon the other productions of the earth. Almost every tree, after its decay, gives origin to a particular species of mushroom. They are often seen as small as pins, with little heads resembling red and yellow beads, growing like a forest under the moist protection of some broad-leaved shrubbery. Over the surface of all accumulations of decayed vegetable matter they are seen spreading out their umbrellas and lifting up their heads, often springing up suddenly, as if by enchantment. But they are short-lived, and soon perish if the light of the sun is admitted into their shady haunts.

Thus far have I endeavored to call attention to the flowerless plants, not designing to treat of them in a scientific manner. I have said nothing, therefore, of the characeæ and the equisetums, lest I make useless repetitions of remarks which are necessarily of a general character. Whoever will take pains to examine these plants will discover an inexhaustible variety in their forms, their modes of growth, and their fructification. Hence those botanists who have given particular attention to this class of vegetation have been noted for the enthusiasm with which they pursued their researches. I have never been initiated into the mysteries of their life, growth, and continuance. I treat of them only as they serve to add beauty to a little nook in the garden, to a dripping rock, or to a solitary dell in the wildwood. The more we study them, the more are we charmed with their singularity and elegance.

Thus, over all her productions has Nature spread the charms of beautiful forms and tints, from the humblest mushroom that grows upon the decayed stump of a tree, or the lichen that hangs in drapery from its living branches, to the lofty tree itself that rears its head among the clouds. It is not in all cases those objects which are most attractive to a superficial observation that furnish the most delight to a scrutinizing mind. The greatest beauties of Nature are hidden from vulgar sight, as if purposely reserved to reward the efforts of those who, with minds devoted to truth, pursue their researches in the great temple of science.

# DROUGHT.

It is an interesting employment to watch the progress of a drought from its commencement, and to witness the efforts of nature to resist its effects and to guard the tender plants from injury. By carefully noting all its phenomena, we may arrive at a knowledge of its causes, which are undoubtedly, in one way or another, connected with the clearing of the forests, and we may learn the means by which we may secure our crops from its ravages, by certain appliances or particular modes of tillage. The drought that visited us in the summer and autumn of 1854, on account of its extraordinary severity and duration, afforded a study for the observer of nature, such as but few generations can witness; and it has led to much speculation concerning the means which may be used to save the country from the frequent recurrence of such an evil. I am but a speculative and superficial observer of these phenomena, having entered only the vestibule of the temple of science. From this I endeavor to take as wide a view as possible of Nature and her works, humbly seeking every opportunity to gain access to the inner temple, from whose windows I may behold a wider prospect, and trace the relations of things which seem now to have no mutual dependence. Many important laws are discovered by correctly noting superficial appearances; and if we trace the connections between all the phenomena that attend one of these periods of drought, we may acquire many points of information that would be valuable both to science and agriculture.

The first symptoms of drought are manifest in the wilting of the grasses, and other rough-leaved and fibrous-rooted plants. Of all perennials, the grasses are the least able to bear continued heat and drought; hence the almost entire absence of this tribe of the vegetable kingdom in the tropics, and their scarcity in all latitudes below the temperate zone. Almost at the same time with the grasses the tender annuals begin to wilt and droop in the gardens. Among our common weeds, the Roman wormwood, the goosefoot, the mustard, and the wild radish feel its effects at an early period. Their leaves become drawn up, they gradually lose their verdure and freshness, and do not increase in growth. Of the annuals, those suffer the least which have a succulent leaf and stem, like the portulacca and the sedum. All the rough-leaved species, like the aster and the hibiscus, are among the first plants that suffer and perish.

As the drought proceeds, the grass fields in the uplands, where the soil is thin and meagre, become dry and yellow; and the clover, the whiteweed, and the saxifrage are made conspicuous by their green tufts, which retain their verdure after the grasses are completely seared. At the same time the foliage of young trees loses its lustre, and is often partially tinted by premature ripeness and decay. The unripened fruits drop constantly from the trees, and the new foliage that is put forth is pale, as if it had suffered from a deficiency of light. The fruits of the season ripen before they have attained their fulness; the whortleberries are withered and dried like pepper, and their foliage is rolled up and crisp. The air contains no moisture, and the hygrometrical lichens upon the rocks and hills are crusty, and break and crumble under our feet, even after sunset.

The lowlands begin to suffer after the uplands seem to be past redemption. Rivers have shrunk to rivulets, and widespread lakes to drear morasses, encircled by a blackened margin, which is exposed by the receding of the waters. Shallow ponds are completely dried, and the fishes are dead and stiffened in the marl which has been baked in the sun. The aquatic plants lift up their long, blackened stems out of the mud, showing the former height of the water, that has sunken away from them. We look round in vain for the usual wild-flowers of the season; they remain blighted and stationary in their growth, and refuse to put forth their blossoms. The whole landscape wears the aspect of a desert; for even the dews have fled, and the evening air is dry and sultry.

The sallow hues of autumn rest upon the brows of summer, like the paleness and wrinkles of age upon a crew that are perishing with starvation and thirst. We find no wet places in the meadows, and even the brooksides can hardly be traced by their greener vegetation. The forest-trees at last begin to suffer; and on the wooded hill-tops we see here and there a group of trees completely browned or blackened in their foliage, some being dead, and others having gone into a state of premature hibernation. The animals suffer no less than the vegetables on which they are dependent. The birds are languid and restless, and do not sing with their accustomed spirit. The squirrels make longer journeys in search of their food, and the hare finds it difficult to obtain sustenance from the dried and tasteless herbs and clover. The chirping insects are dumb and motionless for the want of food; for every tender herb is sear and dry; and multitudes of creatures are hourly perishing with famine.

At this time man watches anxiously for the weather-signs, looking often up to heaven for some kind assurances of relief; but there is no truth in any of the usual omens. The tree-toads from the neighboring orchard — the weather's faithful augurs — by an occasional feeble croaking give false promises of change. The western clouds diminish as they rise; for the fountains of heaven are dried up, and cannot supply them with moisture. The black and threatening clouds that before sunset darken the horizon with the signs of approaching rain are deceitful, and false is every beautiful signal which is hung out amid the splendors of declining day. There is no truth in any sign that appears in the heavens or on the earth.

The arrangements of the clouds do not differ essentially from such as appear in ordinary seasons; but the hue of the heavens is less brilliant, and the tints of yellow and bronze predominate over those of crimson and vermilion. The clouds invariably dissolve soon after sunset, like the steam from boiling water in a clear atmosphere. This dissolution is one of the unfailing accompaniments of drought in all seasons. While the sun is up, there is sufficient evaporation to produce clouds, which continue to increase as long as this moisture is raised by the sun's heat, because the dry region of air above does not absorb them so rapidly as they are produced from below. At the sun's decline all evaporation from the earth and the water ceases. The superfluous moisture of the lower strata of the atmosphere is then precipitated to the earth in the form of dew, and the clouds, which were formed in the daytime, are rapidly absorbed into the dry atmospheric region above them. This process is carried on, day after day, until the whole atmosphere has become saturated, after which rain must soon follow. The first symptoms of this saturation are the continuance of the clouds without any diminution of their bulk after sunset.

The majority of observers have probably witnessed this process of evaporation of the clouds during a dry spell, in the afternoon, — a phenomenon which undoubtedly contributed to give rise to the old saying that "all signs fail in a dry time." The clouds darken and gather together as usual before a thunder-shower. They rise slowly, and sometimes, though but seldom, a slight rumbling of thunder may be heard at a distance. As they ascend above the horizon, their substance becomes perceptibly thinner and more transparent: the fragments that are broken off from their summits dissolve into air, and the clouds will have entirely disappeared before they have risen twenty degrees into the heavens. Not unfrequently a cloud continues to ascend during the prevalence of a strong wind that bears it along so rapidly as to give it no time to dissolve. As we watch its progress with gladness and expectation, we soon observe beneath its dark mass a gleam as bright as gold. The trees and herbs are bent by a brisk and sudden gale; a storm of dust conceals the landscape from sight; a few heavy drops, amidst the din of the elements, splash on the dusty streets, and all is over.

The weather during a period of drought in summer is always even and warm. Any sudden or extreme changes of temperature must necessarily produce rain. Hence fair and serene days are the usual accompaniments of drought. But vegetation is so greatly seared and deprived of its verdure, and all animated things are so listless and silent, that there is but little pleasure in a prospect, except of the ocean and the heavens. It is then delightful to witness the movements of the water-birds, that seem not to share the afflictions of other creatures; and it is refreshing to observe the luxuriance of the marine plants, and to feel the damp and invigorating influence of the sea-breezes that come laden with moisture, and afford a pleasing anticipation of the blessing that must erelong spring from this great reservoir of waters and ultimate source of all the terrestrial gifts of Nature.

# JULY.

THE month of balmy breezes and interminable verdure has given place to one of parching heat and sunshine, which has seared the verdant brows of the hills, and driven away the vernal flowers that crowned their summits. They have fled from the uplands to escape the heat and drought, and have sought shelter in wet places or under the damp shade of woods. Many of the rivulets that gave animation to the prospect in the spring are now marked only by a narrow channel, filled with a luxuriant growth of herbs, that follow its winding course along the plain; and the shallow pools that watered the early cowslips are turned into meads of waving herbage. Millions of bright flowers are nodding their heads over the tall grass, but we scarcely heed them, for they seem like the haughty usurpers of the reign of the meeker flowers of spring. The cattle have taken shelter under the trees to escape the hot beams of the sun, and many may be seen standing in pools or the margins of ponds for refreshment and protection from insects. All animated nature is indulging a languid repose, and the feeble gales hardly shake the leaves of aspen-trees as they pass by them, faint and exhausted with the sultry heats of July.

As June was the month of music and flowers, July is the harvest month of the early fruits; and, though the poet might prefer the former, the present offers the most attractions to the epicure. Strawberries, that gem the meads, and raspberry-bushes that embroider the stone-walls and fences, hang out their ripe, red clusters of berries where the wild-rose and the elder-flower scent the air with their fragrance. The rocks and precipices, so lately crowned with flowers, are festooned with thimbleberries, that spring out in tufts from the mossy crevices half covered with green, umbrageous ferns. There is no spot so barren that it is not covered with something that is beautiful to the sight or grateful to the sense. The little pearly flowers that hung in profusion from the low blueberry-bushes, whose beauty and fragrance we so lately admired, are transformed into azure fruits, that rival the flowers in elegance. Nature would convert us all into epicures by changing into agreeable fruits those beautiful things we contemplated so lately with a tender sentiment allied to that of love. Summer is surely the season of epicurism, as spring is that of the luxury of sentiment. Nature has now bountifully provided for every sense. The trees that afford a pleasant shade are surrounded with an undergrowth of fruitful shrubs, and the winds that fan the brow are laden with odors gathered from beds of roses, azaleas, and honeysuckles. Goldfinches and humming-birds peep down upon us, as they flit among the branches of the trees, and butterflies settle upon the flowers and charm our eyes with their gorgeous colors. In the pastures the red lilies have appeared, and young children who go out into the fields to gather these simple luxuries, after filling their baskets with fruit, crown their arms with bouquets of lilies, laurels, and honeysuckles, rejoicing over their beauty during the happiest, as it is the most simple and natural, period of their lives.

There is not a more agreeable recreation at the present season than a boat-excursion upon a wood-skirted pond, when its alluvial borders are brightly spangled with water-lilies, and the air is full of delicate incense from their sweet-scented flowers. The plover may be seen gliding with nimble feet upon the broad leaves that float on the surface of the waters, so lightly as hardly to impress a dimple on the glossy sheen; and multitudes of fishes are gambolling among their long stems in the clear depths below. Among the fragrant white lilies are interspersed the more curious though less delicate flowers of the yellow lily; and in clusters here and there upon the shore, where the turf is dank and tremulous, the purple sarracenias bow their heads over lands that have never felt the plough. The alders and birches cast a beautiful shade upon the mirrored border of the lake, the birds are singing melodiously among their branches, and clusters of ripe raspberries overhang the banks as we sail along their shelvy sides.

But we listen in vain on our rural excursions for the songs of multitudes of birds that were tuneful a few weeks since. The chattering bobolink, merriest bird of June, has become silent; he will soon doff his black coat and yellow epaulettes, and put on the russet garb of winter. His voice is heard no more in concert with the general anthem of Nature. He has become silent with all his merry kindred, and, instead of the lively notes poured out so merrily for the space of two months, we hear only a plaintive chirping, as the birds wander about the fields in scattered parties, no longer employed in the cares of wedded life. But there are several of our warblers that still remain tuneful. The little wood-sparrow sings more loudly than ever, the vireo and wren still enliven the gardens, and the hermit-thrush daily utters his liquid strains from his deep sylvan retreat upon the wooded hills.

In the place of the birds myriads of chirping insects pour forth during the heat of the day a continual din of merry voices. Day by day are they stringing their harps anew, and leading out a fresh host of musicians, making ready to gladden the autumn with the fulness of their songs. At intervals during the hottest of the weather, we hear the peculiar spinning notes of the harvest-fly, a species of locust, beginning low and with a gradual swell, increasing in loudness for a few seconds, then slowly dying away into silence. To my mind these sounds are vivid remembrancers of the pleasures and languishment of noonday, of cool shades apart from sultry heats, of repose beneath embowering canopies of willows, or grateful repasts of fruits in the summer orchard.

The season of haymaking has arrived, the mowers are busy in their occupation, and the whetting of the scythe blends harmoniously with the sounds of animated nature. The air is filled with the fragrance of new-mown hay, the dying incense-offering of the troops of flowers that perish beneath the fatal scythe. Many are the delightful remembrances connected with haymaking to those who have spent their youth in the country. In moderate summer weather there is no more delightful occupation. Every toil is pleasant that leads us into green fields and fills the mind with the cheerfulness of all living things.

But summer, with all its delightful occasions of joy and rejoicing, is in one respect the most melancholy season of the year. We are now the constant witnesses of some regretful change in the aspect of nature, reminding us of the fate of all things and the transitoriness of existence. Every morning sun looks down upon the graves of whole tribes of flowers that were but yesterday the pride and glory of the fields. Day by day as I pursue my walks, while rejoicing at the discovery of some new and beautiful visitant of the meads, I am suddenly affected with sorrow upon looking around in vain for the little companion of my former excursions, now drooping and faded and breathing its last breath of fragrance into the air.

I am then reminded of early friends who are no longer with the living; who were cut down, one by one, like the flowers, leaving their places to be supplied with new friends, perhaps equally lovely and worthy of our affections, but whose even greater loveliness and worth will never comfort us for the loss of those who have departed. Like flowers, they smiled upon us for a brief season, and, like flowers, they perished after remaining with us but to teach us how to love and how to mourn. The birds likewise sojourn with us only long enough to remind us of the joy of their presence and to afford us an occasion of sorrow when they leave us. We have hardly grown familiar with their songs ere they become silent and prepare for their annual migration. They are like those agreeable companions among our friends who are ever roaming about the world on errands of duty or pleasure, and who only divide with us that pleasant intercourse which they share with other friendly circles in different parts of the earth.

It is now midsummer. Already do we perceive the lengthening of the nights and the shortening of the sun's diurnal orbit. We are reminded by the first observation of this change that summer is rapidly passing away; and we think upon it with a painful sense of the mutability of the seasons. But let us not lament that Nature has ordained these alternations; for though there is no change that does not bring with it some lingering sorrows over the past, yet may it not be that these vicissitudes are the true sources of that happiness which we attribute only to the immediate causes of pleasure? Every month, while it sadly reminds us of the departed joys and beauties of the last, brings with it a recompense in bounties and blessings which the preceding month could not afford. While rejoicing, therefore, amid the voluptuous delights of summer, we will not regret that we cannot live forever among enervating luxuries. With the aid of temperance and virtue, all seasons as they come may be made equal sources of enjoyment. And may it not be that life itself is but a season in the revolving year of eternity, the vernal season of our immortality, that leads not round and round in a circle, but onward, in an everlasting progression, to greater goodness and greater bliss, until the virtues we now cherish have ripened into eternal felicity?

# PROTECTION OF BIRDS.

THE presence of birds as companions of a home in the country is desirable to all, next to woods, flowers, green fields, and pleasant prospects. Without birds, the landscape, if not wanting in beauty, would lack something which is necessary to the happiness of all men who are above a savage or a boor. Indeed, it is highly probable that Nature owes more to the lively motions, songs, and chattering of the feathered race for the benign effects of her charms, than to any other single accompaniment of natural scenery. They are so intimately associated with all that is delightful in field and forest, with our early walks in the morning, our rest at noonday, and our meditations at sunset, with the trees that spread their branches over our heads, and the lively verdure at our feet, that it is difficult to think of one apart from the others. Through the voices of birds Nature may be said to speak to us, and without them she would be a dumb companion whose beauty would hardly be felt.

Both from our regard for their utility to agriculture and for their pleasant companionship with man, we have thousands of motives for protecting the birds. Very little attention has been paid to this subject. A few laws have been made for their preservation; but they have seldom been enforced. I believe the farmer would promote his own thrift by extending a watchful care over all families of birds, but the smaller species are the most useful and delightful. It seems as if Nature had given them beauty of plumage and endowed them with song, that man by their attractions might be induced to preserve a race of creatures so valuable to his interest.

There are two ways of preserving the birds: we may avoid destroying them, and we may promote the growth of certain trees, shrubs, and plants that afford them shelter and subsistence. The familiar birds that live in our gardens and orchards will multiply in proportion as the forests are cleared and the land devoted to tillage, if the clearing does not amount to baldness. To this class belong many of our sparrows, the robin, the bobolink, — indeed, all our familiar species. The solitary birds that inhabit the pasture and forest would probably be exterminated by the same operations that would increase the number of robins and sparrows. It is no less necessary to keep the birds for the preservation of the forests than to keep the forests for the preservation of the birds.

To insure the protection of all species, there must be a certain proportion of thicket and wildwood. The little wood-sparrow seldom frequents our villages, unless they are closely surrounded by woods. Yet this bird lives and breeds in the open field. He frequents the pastures which are overgrown with wild shrubbery and its accompaniment of vines, mosses, and ferns. He is always found in the whortleberry field, and probably makes an occasional repast on its fruits in their season. He builds his nest on the ground or on a mossy knoll protected by a thicket. All birds are attached to grounds which are covered with particular kinds of plants and shrubbery that sustain their favorite insect food. If we destroy this kind of vegetation, we drive away the species that are chiefly attached to it from our vicinity, to seek their natural habitats. We may thus account for the silence that pervades the locality of many admired country-seats; for with regard to the wants of our familiar birds it is often that trimming and cultivation are carried to a pernicious extreme.

There will be no danger for many years to come that our lands will be so thoroughly stripped of their native growth of herbs, trees, and shrubs as to leave the birds without their natural shelter in some places. The danger that awaits them is that they may be driven out of particular localities, and the inhabitants thereby deprived of the presence of many interesting songsters. Wherever the native species are abundant, we find a considerable proportion of cultivated land, numerous orchards, extensive fields of grass and grain, interspersed with fragments of forest or wildwood, well provided with watercourses. Where these conditions are present, the familiar birds will be numerous if they are not destroyed. If these cultivated lands lie in the vicinity of pastures abounding in thickets and wild shrubbery, fragments of wood and their indigenous undergrowth, we may then hear occasionally the notes of the solitary birds, many of which are superior in song. Wild shrubbery and its carpet of vines and mosses form the conditions that are necessary for the preservation of these less familiar species.

The shrubs that bear fruit are the most useful to the birds, especially as they are infested by more insects than other kinds. The vaccinium, the viburnum, the cornel, the elder, the celastrus, and the small cherries are abundant where there is a goodly number of the less familiar birds. If we clear our woods of their undergrowth and convert them into parks, we do in the same proportion diminish the numbers of many species. No such clearing as this is favorable to any of the feathered race. But the clearing and cultivation of the land outside of the woods, if it be done rudely, leaving bushes on all barren knolls and elevations, is beneficial to all kinds of birds by increasing the quantity of insect food in the soil. A nice man at the head of a farm would do more to prevent the multiplication of birds, than a dozen striplings with their guns. The removal of this miscellaneous undergrowth and border shrubbery would as effectually banish the red-thrush, the catbird, and the smaller thrushes, as we should extirpate the squirrels by destroying all the nut-bearing trees and shrubs.

A smooth-shaven green is delightful to the eye at all times; but lawn is a luxury that is obtained at the expense of the familiar birds that nestle upon the ground. The song-sparrows build their nests in the most frequented places, if they are not liable to be disturbed. Not a rod from our dwelling-house these little birds may have their nests, if the right conditions are there. They are often built on the side of a mound overrun by blackberry-vines and wild rose-bushes. He who would entice them to breed in his enclosures must not, for the preservation of a foolish kind of neatness, eradicate the native shrubs and vines as useless weeds.

Clipped hedge-rows, which have been recommended as nurseries of birds, are checks to their multiplication. A hedge-row cannot be "properly" maintained without keeping the soil about its roots clear of grass and wild herbage, which are needful to the birds. It is only a neglected hedge-row that is useful to them, or a spontaneous growth of bushes and briers, such as constitutes one of the picturesque attractions of a New England stone-wall. We seldom see one that is not covered on each side with roses, brambles, spirea, viburnum, and other native vines and shrubs, so that in some of our open fields the stone-walls, with their accompaniments, are the most attractive objects in the landscape. Along their borders Nature calls out, in their season, the anemone, the violet, the cranesbill, the bellwort, the convolvulus, and many other flowers of exceeding beauty, while the rest of the field is devoted to tillage.

The "nice man" who undertakes farming will grudge

Nature this narrow strip on each side of his fences, though she never fails to cover it with beauty. He considers it an offence against neatness and order to allow Nature these simple privileges, and employs his hired men to keep down every plant that dares to peep out from the fence-border without a license from the owner. Such a miscellaneous hedge-row would constitute a perfect aviary of singing-birds, and the benefits they would confer upon the farmer by ridding his lands of noxious insects would amply compensate him for the space left unimproved. Then might we hear the notes of the wood-thrush and the red-mavis in the very centre of our villages, and hundreds of small birds of different species would cheer us by their songs where at present only a solitary individual is to be heard.

From the earliest times it has been customary to encourage the multiplication of swallows by the erection of bird-houses in gardens and enclosures. Even the Indians furnished a hospitable retreat for the purple martin by fixing hollow gourds and calabashes upon the branches of trees near their cabins. It is generally believed that this active little bird is capable of driving away hawks and crows from its vicinity by repeated annoyances. The custom of supplying martins with a shelter has of late grown into disuse. The wren and the bluebird may be encouraged by similar accommodations. But as these two species are not social in their habits of building, like the martin, a separate box must be supplied for each pair of birds. The wren is an indefatigable destroyer of insects and one of the most interesting of our familiar songsters. The bluebird, which is not less familiar, is delighted with the hollow branch of an old tree in an orchard, but is equally well satisfied with a box.

# BIRDS OF THE PASTURE AND FOREST.

## III.

### THE CUCKOO.

Our native Cuckoos have not the free-love instinct of the European Cuckoo; and Daines Barrington would have been delighted to quote their good parental habits as an argument in his special plea for the European bird, whom he considered the victim of slander. The Cowbird is our Cuckoo in the moral acceptation of the term. The American Cuckoo is attached to its offspring in a remarkable degree, and rears them with all the fidelity of the most devoted parents. In my boyhood, the two severest fights I had with birds on approaching their nests were once when I examined the nest of a Bluejay, and again when I examined one belonging to a Cuckoo. The young Cuckoos were equally savage when I attempted to handle them. Yet this bird bears the reputation of cowardice.

It is remarkable that the American Cuckoo, though a faithful and devoted parent, should have certain peculiar habits connected with laying and hatching, that bear some evidence that the European and American species have a common derivation. The habit of the European bird of dropping its eggs into other birds' nests is probably connected with continued laying, extended to a greater length of time than with other birds. The same fertility has been observed in the American Cuckoos. Mr. Audubon mentions the peculiar habit of these birds of laying fresh eggs and hatching them successively. Thus

140

it would seem that the last-laid eggs were hatched by the involuntary brooding of the young which had not left the nest. Dr. Brewer has "repeatedly found in a nest three young and two eggs, one of the latter nearly fresh, one with the embryo half developed, while of the young birds, one would be just out of the shell, one half fledged, and one just ready to fly. Subsequent observations in successive seasons led to the conviction that both the Yellow-billed and the Black-billed Cuckoo share in these peculiarities, and that it is a general but not universal practice."

Dr. Brewer mentions an interesting fact that evinces the strong attachment of the Cuckoo to its offspring. Speaking of the Black-billed Cuckoo, he says: "Both parents are assiduous in the duties of incubation and in supplying food to each other and their offspring. In one instance where the female had been shot by a thoughtless boy, as she flew from the nest, the male bird successfully devoted himself to the solitary duty of rearing the brood of five. At the time of the death of the female, the nest contained two eggs and three young birds. The writer was present when the bird was shot, and was unable to interfere in season to prevent it. Returning to the spot not long afterwards, he found the widowed male sitting upon the nest, and so unwilling to leave it as almost to permit himself to be captured by the hand. His fidelity and his entreaties were not disregarded. This nest, eggs, and young were left undisturbed; and as they were visited from time to time, the young nestlings were found to thrive under his vigilant care. The eggs were hatched out, and in time the whole five were reared in safety."

The Cuckoo is an early visitor. His voice is often heard before the first of May, proclaiming that "the spring is coming in," like his congener in England, who has always been regarded as the harbinger of that season.

His note is not strictly musical, yet we all listen to the first sound of his voice with as much pleasure as to that of the Bluebird or Song-Sparrow. I have not met a person who was not delighted to hear it. It may be called, figuratively, one of the picturesque sounds in Nature, reminding us of the resurrection of the long-hidden charms of the season. The Cuckoo is swift in his flight, which resembles that of a Dove so much that I have often mistaken them. In plumage and general shape this bird is like the Red-Thrush, with some mixture of olive.

### THE COWBIRD.

Young nest-hunters, who are persistent in their enterprises, and who pursue their occupation partly from rational curiosity and not from mere wantonness, are often surprised on finding in the nest of some small bird a single egg larger than others in the same nest. In my own days of academic truancy, I found this superfluous egg most frequently in Sparrows' nests. It was not until I had made a large collection of eggs that I discovered the parentage of the odd ones. These eggs were generally speckled; but I occasionally found a large bluish egg among others of the same color, and supposed they must contain two yolks, save that birds in a wild state seldom produce such monstrosities. Can it be that the American Cuckoo occasionally follows the instincts of his European congener? In each case I considered the spurious eggs as lawful plunder, since they were an imposition practised upon the owner of the nest either by some unknown bird or by the Cowbird, a member of a family which are too aristocratic to rear their own offspring. But as a politician of the speculative class I feel a peculiar interest in the Cowbird, as affording me an opportunity of understanding the system of free love, as exemplified in the habits of this species.

The Cowbird has no song. Nature seldom furnishes any creature with an instinct which would be of no service to the species. What occasion has the Cowbird for a song, — a bird that neither wooes nor marries, — a bird that would not sing lullabies to its own young; that cares no more for one female than for another, and whose indifference is perfectly reciprocated? As well might a poet write Petrarchian sonnets who was never in love; or a practical plodder write amatory songs, who asks the members of a church whom he shall marry. There is nothing romantic in this bird's character. His love is a mere gravitation. Nature, despising his habits, has not even arrayed him in attractive plumage; for why should he have beauty when his whole species are without the sentiment that could appreciate it? The Cowbirds are the free-love party among the feathered tribes, — the party also of communism, who would leave their offspring in others' hands, that they may have leisure for æsthetic culture.

"This species," says Dr. Brewer, "is at all times gregarious and polygamous, never mating and never exhibiting any signs of either conjugal or parental affection. Like the Cuckoos of Europe, our Cow-Blackbird never constructs a nest of her own, and never hatches out or attempts to rear her own offspring, but imposes her eggs upon other birds; and most of them, either unconscious of the imposition or unable to rid themselves of the alien, sit upon and hatch the stranger, and in so doing virtually destroy their own offspring; for the eggs of the Cowbird are the first hatched, usually two days before the others. The nursling is much larger in size, filling up a large portion of the nest, and is insatiable in appetite, always clamoring to be fed, and receiving by far the larger share of the food brought to her nest; its foster companions, either starved or stifled, soon die, and their dead bodies are removed, it is supposed, by the parents. They are never found near the nest, as they would be if the young Cow-Blackbird expelled them as does the Cuckoo; indeed, Mr. Nuttall has seen parent birds removing the dead young to a distance from the nest and there dropping them."

### THE REDWING-BLACKBIRD.

In early spring no sounds attract so much attention as the unmusical notes of the Redwing-Blackbird coming to our ears from every wooded meadow. A sort of *chip-chip churee*, mixed with many other confused and some guttural sounds, forms this remarkable chorus, which seems to be a universal chattering, hardly to be considered a song. Most of the notes are sharp, and in none could I ever detect anything like musical intonation. Sometimes they seem to chant in concert with the little piping frogs, though the sounds made by the latter are by far the most musical. Indeed, the Redwing-Blackbird never sings, though we frequently hear from a solitary individual the sound of *chip-churee*.

This bird, as well as the Cowbird, is a free-lover, though the females have not yet declared their rights, and their communistic prejudices are not sufficient to cause them to refuse to rear and educate their offspring. In early April assemblages of Redwings, perched upon trees standing in wet grounds, constantly chatter in merry riot, while the bright scarlet epauletted males strive to recommend themselves by music, like some awkward youth who serenades his mistress with a jewsharp. These notes seem to spring from a fulness of joy upon returning to their native swamps. The Redwings undoubtedly mate, though there is plainly no jealousy among them. Like the Otaheitans, a flock of birds has a flock of wives, the true wife being recognized above the others only while rearing their young. In this respect they differ from the gallinaceous birds, who resolutely demand exclusive possession of all the females and establish their right by might. They fight until the conqueror is left to be the sultan of the flock.

The nests of the Redwing are always suspended upon a bush or a tuft of reeds in a half-inundated meadow. I have frequently found them in a button-bush, surrounded by water; but they are also suspended from the perpendicular stalks of cat-tails, which encircle the nests, bound to them by the leaves of the same plant or any other fibrous material which is near at hand. The Redwing displays almost as much dexterity as the Baltimore Oriole in the construction of its nest, which is always firmly woven so that it is not easily detached from its position. It rears but one brood in a season. The eggs have a whitish ground tinged slightly with blue, and mottled with dark purple blotches irregularly distributed. The Redwings are resolute defenders of their nest and young, both parents manifesting equal anxiety and courage.

Like all our most useful birds, the Redwings are very mischievous, consuming Indian corn while it is in the milk, and thus doing an incalculable amount of damage, especially at the South, where the species assemble in countless flocks. Alexander Wilson has seen them so numerous in Virginia during the month of January, as to resemble an immense black cloud. When they settled upon a meadow their united voices made a sound which, heard at a distance, was sublime; and when they all rose together upon the wing, the noise was like distant thunder. He took particular notice of the glitter of their epaulets, flashing from thousands of wings from this vast assemblage. At the North they are seldom numerous enough to do any extensive damage, and they are such indefatigable hunters of all those grubs that are concealed beneath the surface of the ground, that they probably compensate in this way for all the mischief they perform.

### THE PURPLE GRACKLE.

High up in the pines or firs that constitute a grove outside of any of our villages, in the latter part of April, small flocks of Purple Grackles may be seen gathered together like Rooks, and making the whole neighborhood resound with their garrulity. They are not very shy birds, seeming hardly conscious of the enmity with which they are regarded by the villagers near whose habitations they congregate. They become every year more numerous and familiar, their numbers increasing with the extension of the area of tillage. In no way is the truth of the Malthusian theory more clearly proved or more plainly illustrated than in the habits of certain species of birds. They will increase in spite of our persistent efforts to exterminate them, unless we cut down our woods and thickets to deprive them of a shelter and a home. A single model farmer or landscape-gardener may do more in the way of their extermination, by keeping his grounds nice, and clear of undergrowth, than twenty mischievous boys with guns or a dozen avaricious farmers with their nets. Birds that, like the Robin and the Grackle, consume all sorts of insects they can find upon the ground, will increase with their supply of insect food. If we wish to stop their multiplication, we must bury every fertilizer six feet deep.

The Grackles are intelligent birds, and, though apparently not very shy, they are wise enough to build their nests in the tops of tall trees which are difficult of access, choosing an evergreen for this purpose, that they may be more safely concealed. These birds have been known to build sometimes in the hollows of trees; likewise inside of the spire of a church and in martin-houses. Indeed, Mr. S. P. Fowler thinks that as human population increases, the Grackles are gradually assuming the habits of the English Rooks. Like the Rook, they are naturally gregarious, and as the area of agriculture is expanded, and woods afford birds less protection than formerly, they are disposed to seek artificial shelter in the vicinity of towns, that they may feed upon insect food, which in these localities is very abundant.

The Purple Grackle has, upon examination, very beautiful plumage; for its black feathers are full of various tints, changeable, according as the light falls upon them, into violet, purple, blue, and green. We see, however, nearly all the same varying shades in the plumage of the common Cock, when it is black. They are said to consume so much corn as to seriously injure the crop wherever they exist in large numbers. Still they are so useful as to deserve not only protection, but encouragement, and groves in which they can nestle without disturbance should be saved for them.

Like the Redwing, they assemble in large flocks in the Southern States. According to Wilson, the magnitude of their assemblages can hardly be described. In Virginia he witnessed one of these myriad flocks settled on the banks of the Roanoke. When they arose at his approach, the noise of their wings was like distant thunder, and they completely hid from sight the fields over which they passed by the blackness of their multitudinous flocks. He thought the assemblage might contain hundreds of thousands. The depredations of such immense flocks upon the Indian-corn crop must be incalculable, since they are known to attack it in all stages of its growth, beginning as soon as it is planted.

In New England they remain only during the breeding-season, when it is a well-established fact that their whole diet consists of worms and insects. Good observers who have watched them here testify to the truth of this assertion. They do, in fact, consume but little corn or grain at any season, save when they cannot find a sufficient supply of insect food. When associated in such vast flocks as described by Wilson, they are necessarily granivorous.

### THE QUAIL.

I have not yet seen any good reason for denying that the Quail is a Quail; nor can I understand why, in the new classifications of birds, the marks that formerly characterized species are now used to characterize genera. Let us pursue the same philosophical rule to its final results, and we shall arrive at the discovery that the different varieties of the common fowl constitute so many genera, and that the black and the white and the Seebright Bantams, for example, are species of the genus *Galliparvus*. But the Quail, whether it be itself or another bird, is now a rare inhabitant of New England. Thousands of its species were destroyed by the deep snows of the winter of 1856-57, and again by the cold winter of 1867-68. Indeed, every winter destroys great numbers of them. And as the Quail does not migrate, and never wanders any great distance from its birthplace, I cannot understand why its species could ever have been numerous so far north as the New England States, unless the vast numbers rendered it impossible for any accident of Nature to destroy so many that there should not be multitudes left. But since the white man came, the gun, the snare, and the winters united have nearly extirpated the whole race.

For many years past I have seldom heard the musical voice of the Quail. Seldom is the haymaker in these days reminded of the approach of showers by his proclamation of "More wet" from some adjoining fence. Not that the few that remain are no longer prophets, but they have become timid from the persecutions they have suffered, and have ceased to prophesy in the vicinity of the farm. Neither does the Quail any longer make known his presence to his mate by saying in musical tones, "Here's Bob White." He knows too well that this would lead to his discovery and death. Man, too short-sighted to understand his own selfish advantage in protecting the bird, and too avaricious to let pass the opportunity of buying a feast with a few cheap charges of powder and shot, will give him no peace.

A female Quail, leading her little brood under the shelter of pines to escape the notice of those who have intruded into her presence, is one of the most interesting sights in animated nature. The rapidity with which the young make their escape to some hiding-place in the grass or among the bushes, and the anxiety displayed by the mother, cannot fail to awaken our sympathy. If we sit still in ambush and watch for them, the mother, no longer aware of our presence, gives her cheerful call-note, when they all suddenly reappear and follow her, as chickens follow the hen. Their timidity and their expertness in wending their way through the thicket and then out on the open land, and their nimble motions as they forage in the pasture for grubs and insects, are an ample reward to any sympathetic observer for long and patient watching.

The destruction of this useful and interesting species by our winter snows is a public calamity; and nothing, it seems to me, can mitigate the evil save the building of artificial shelters, strewing around them some sort of grain to prevent their wandering far away from them. Our farmers have not sufficiently considered the advantages they might derive from this semi-domestication of the Quail and some other species that winter with us. Even if this protection were offered them only that their surplus might be used to grace our tables, it would be found a profitable enterprise.

### THE RUFFED GROUSE.

In May, if we were to wander into an extensive wood which is not a swamp, at a sufficient distance from any village tavern, we should probably hear the drumming of the Partridge. This peculiar sound is heard early in the morning and late in the evening, becoming more frequent and persistent as the breeding-season advances. It is made by the male, and is unlike any other sound I ever heard. I cannot compare it to the rumbling of distant thunder, as some do, because the sounds of thunder are irregular, while the strokes of the Partridge's wings are perfectly timed, and increase in rapidity as they decrease in loudness, until they die away in a faint, fluttering vibration.

I think those observers are mistaken who believe this drumming to be made by striking or flapping his wings against his sides or against the log where he is standing. Samuels says: "The bird resorts to a fallen trunk of a tree or log, and while strutting like a male Turkey, beats his wings against his sides and the log with considerable force. It commences very slowly, and after a few strokes gradually increases in velocity, and terminates with a rolling beat very similar to the roll of a drum." Dr. Brewer describes the sound as produced in the same manner, and this seems to be the universal opinion. On the contrary, the bird produces this sound by striking the shoulders of his wings together over his back, as the common Cock frequently does before he crows, and as the male Pigeon does when after dalliance with his mate he flies out exultingly a short distance from his perch. It is very difficult to obtain sight of the bird while he is drumming, and then we cannot venture near enough to see his motions very distinctly. But whenever I have gained sight of one in the act of drumming, he seemed to me to elevate his wings and strike them together over his back, increasing the rapidity of these strokes, until the last was nothing more than the sound produced upon the air by the rapid vibration of the feathers of his wings and tail. A similar vibrating sound is made by the Turkey with his tail-feathers when strutting about the yard among the females.

It seems very improbable that the Grouse has sufficient power to make so much sound by flapping the concave surface of his wings against his downy sides. Birds cannot move their wings with so much force in this direction as in the opposite one; and so long as some uncertainty exists about it, it is the wisest course to reason from analogy, and to conclude that the Partridge makes this sound as similar ones are made by certain domestic birds. Many of our farmers believe that this bird stands on a log and makes the drumming sound by striking the shoulders of his wings against the log. Some think the log must therefore be hollow. But instances are well known where a bird has selected a rock for his drumming-place, when the same sound is produced.

As the flapping of the wings of the common Cock previous to crowing is a mode of expressing defiance, the same may be said of the drumming of the Partridge, who before and after his drumming struts about in the most amusing way, placing himself in many graceful attitudes. All these actions are a part of the ceremony of courtship. They always, therefore, excite the jealousy of other males, who, if sufficiently bold, will immediately attack the drummer. The conqueror draws in his train the greater part of the females, and becomes their favorite.

## SIMPLES AND SIMPLERS.

WHEN chemistry had become elevated to an equal rank with the other exact sciences, physicians, who in the days of alchemy and astrology had dealt almost exclusively in simples, discarded from their practice the greater part of the herbs of the old pharmacopœias, and used in the place of them the more certain and efficacious preparations of the laboratory. The metals, in the various forms of oxides, carbonates, chlorides, sulphurets, and other chemical compositions, were proved to be more decided and commensurable in their action upon the human system than roots and herbs. Chemistry took the place of botany to a great extent in the healing art, and caused a gradual division of the practice and the dispensation of medicine. The apothecary assumed the department of preparing and compounding the drugs used by the physician; and as the medical faculty dropped the general use of simples, the dispensation of them naturally fell into the hands of certain individuals of the female sex. They became the conservators of ancient medical notions that science had rejected, and gradually introduced a sort of domestic practice which is not yet entirely discontinued.

They were, indeed, the traditional followers of the practice of the early physicians, when they were simplers and astrologers, and administered to the wants of those people who believed the herbs of the field to be the only safe remedies for disease. Their botanical knowledge was confined to the mere identification of plants, and to certain ancient classifications of medical herbs made on a somewhat arbitrary principle, and dictated by a love of formal arrangement that distinguished the learned of the Middle Ages. They knew the "Four Great Carminative Hot Seeds, and the Four Lesser Hot Seeds; the Four Cold Seeds, and the Four Lesser Cold Seeds; the Five Opening Roots, and the Five Lesser Opening Roots; the Five Emollient Herbs; the Five Capillary Herbs; the Four Sudorific Woods; the Four Cordial Flowers; the Four Carminative Flowers, and the Four Resolvent Meals." Here was a botanical arrangement of plants precisely like that of the Five Orders of Architecture. Though extremely artificial, it was founded on the real or supposed properties of the plants included in it. Its formality suited the taste and assisted the memory of the simplers. They could understand it, and they were proud of their knowledge, because they derived from it an important consideration in their own village.

There was no danger in trusting one's health to the judgment and mercy of these honest women. They were not guilty, like our modern inventors of patent medicines, of furnishing a powerful drug disguised in a decoction of some popular herb. Their teas, syrups, and fomentations, their lotions, quilts, diet-drinks, and electuaries, were made from the herbs which were specified among their ingredients, and were safe even when injudiciously applied. They dealt in no dangerous remedies; some were only cordial beverages, some were mild emetics, tonics, and refrigerants, and very many of them were entirely inert, but like an amulet soothing to the mind. In the days of our grandmothers, almost every garden contained the herbs of their simple dispensatory; and every neighborhood was graced by a goodly number of housewives who were versed in all details of the administration of them. In these old gardens were mints of every sort, basil, rosemary, fennel, tansy, spikenard, blessed thistle, and saffron. No garden was considered properly furnished if it were wanting in any of the herbs that might be required by the sick of the neighborhood. Flowers cultivated for their beauty were also the occupants of these gardens; roses, in particular, which were as needful in their dispensation as the chief of the cordial herbs.

The mints were held in great esteem by these charitable dames. They paid special attention to spearmint, — regarded as the mint of mints, — the smell of. which was believed to "corroborate the brain and increase and preserve the memory," and it was venerated like one of the holy herbs. Hardly less value was affixed to the basil, once considered a "royal plant," on account of its excellent properties. It is remarkable that in the time of the ancient Romans the basil was believed to possess the power of breeding serpents. Hence, when they sowed the seeds of this plant they bestowed curses upon it, that it might be dispossessed of its nefarious properties by their maledictions. This notion did not descend to the English people. By them and by our simplers it was cherished for its sweet smell, which was "good for the heart and the head"; also for its "seed that cureth the infirmities of the heart, and taketh away sorrowfulness which cometh of melancholy, and maketh a man merry and glad." The sweet-marjoram, which still retains its popularity as a savory herb, was famous in these old gardens, and then known as the celebrated "Dittany of Crete." At present it is not used as a medicine in any form; but the simplers believed it to be efficacious in restoring the sense of smell when it was lost, and it was noted for its vulnerary powers.

Many of the herbs of their dispensatory were formerly dedicated to the Virgin or to some worshipful saint, and were considered holy. Probably in some cases their supposed medical virtues were deduced from their sanctity; in other cases their real virtues may have caused them to be religiously consecrated. It does not appear that sectarian prejudice raised any distrust in their Protestant minds of the merits of a plant which had derived its sanctity from Roman-Catholic usages. Among the early Romans plants were supposed to derive their virtues from some rural deity to whom they were dedicated; and the curative powers of mineral waters were attributed to the nymph who presided over the spring; and those who drank at the fountain worshipped the beautiful goddess from whose divine qualities its virtues emanated. When the heathen world was converted to the religion of Christ, these superstitions changed their character, but were not cast aside. Holy wells and fountains still retained the veneration of men; but their virtues were ascribed to saints, and not to water-nymphs.

A savor of romance still adheres to many of the holy plants, derived from the incidents that led to their consecration. The costmary, an Italian plant not uncommon in our gardens, having a very agreeable aromatic odor and some peculiar balsamic properties, was, on account of the purity of its fragrance, dedicated to the Virgin. In its sensible qualities it unites the balm and the tansy. The blessed thistle, another of the holy herbs, is one of those plants that may be compared to certain good people whose virtues are all of a passive sort, and who are chiefly remarkable for the odor of sanctity that distinguishes them. Some other herbs have won their reputation from their supposed identity with certain plants mentioned in Scripture. There are likewise holy shrubs, as the waybread and the wayfaring-tree, — names highly suggestive and romantic. Others, like the witch-elm and the witch-hazel, are associated with divination and magic. In Great Britain, where the habits of the people are still under traditional influences in a much greater degree than those of the same classes in this country, a profound respect is still paid to the holy herbs; and bands of simplers — believers in the panaceas of the field and garden — still continue their avocation and are in popular repute in many old English towns.

During the infancy of modern science, when theology was mingled with all the exercises of the mind, and when it was believed that everything was created for man's especial use, all plants were supposed, as a doctrine of religious faith, to contain some qualities, discovered or undiscovered, which were intended by Providence for the sustenance, protection, and clothing of man, or for the cure of his diseases. The flowerless plants, now known to be without any curative properties, were then extensively used in medicine, from the pious supposition that, as they are useless for food or for employment in practical arts, they must be intended by Divine Providence for medicines. In that romantic era, pillows were filled with the substance of a kind of moss which was supposed to be useful for procuring sleep. The family of mosses from which this substance was obtained, in accordance with the use made of it, received from the early botanists the name of *hypnum*, from a Greek word that signifies "sleep." This was afterwards combined with other products, such as poppy leaves, wormwood, the petals of the peony, and the flowers of hops, and used for similar purposes in the form of quilts. These substances were placed between two pieces of cotton or linen, and quilted into a cap to be worn on the head for the headache. They were made also in other forms, to be laid upon any part affected with inflammation or nervous pains.

The doctrine of signatures, believed by the whole Christian world in the Middle Ages, was a theory of religious philosophy, and shows the intimate connection existing in that era between theology and medicine. According to this theory every natural substance that possesses any curative power indicates by its external appearance the diseases for which it is a remedy. The partisans of this doctrine affirmed that, since man is the lord of creation, all creatures are designed for his use; and that therefore their properties must be designated by such a character as every one can understand. Hence turmeric, or Indian saffron, which has a brilliant yellow color, indicates thereby its power of curing the jaundice. By the same rule poppies were believed to be a cure for diseases of the head, because both their seeds and flowers form a head. A beautiful flower called *euphrasia*, or eye-bright, resembling a dandelion, with a dark, velvety centre, was used for diseases of the eye, because this dark round centre bears a likeness to the pupil of the eye. In this doctrine we find an anticipation of the homœopathic theory of " like cures like." Nettle-tea in England still continues to be a popular remedy for nettle-rash.

The flowers of saffron, of a bright scarlet color, which are administered in the form of tea for scarlet-fever and other eruptive diseases, derived all their reputation from the homœopathic doctrine of signatures, expressed in the words *similia similibus curantur*. Hence likewise the celebrated botanical cure of hot-drops administered in fevers, on the supposition that a hot disease requires a hot remedy; and the ancient notion that the hair of a mad dog will cure the disease caused by his bite. These analogies have been indefinitely extended. The blood-root is another of the signature plants. Its clusters of delicate white flowers appear in April in damp shady places. It avoids the deep woods and seeks the protection of clumps of trees near a brookside, where the soil is deep, and the situation defended by a natural wall or embankment. Its tuberous root is full of red sap resembling blood. Hence it was considered the natural remedy for all blood diseases. It is seldom used in modern legitimate practice. The liverwort (*Hepatica triloba*), a beautiful early flowering anemone, not uncommon in our woods, was used as a cure for liver complaints, from the resemblance of its leaf, which is lobed, to the folds of the liver, and of its mottled hues of green and purple to the outward colors of the liver. This plant is still in use by our modern simplers.

In the use of the five capillary herbs we trace the influence of the doctrine of signatures. All these herbs were ferns: the hartstongue, black, white, and golden maidenhair, and spleenwort. These plants, when they first appear above the ground, are covered with hairy down. This appearance caused them to be credited with efficacy in improving the growth of the hair, hence named capillary herbs. There are three distinct species of maidenhair in this catalogue,—the black, white, and golden, representing the colors of the human hair in childhood, manhood, and old age. The stems of these beautiful ferns are also nearly as slender as hairs; another signification of their proper medical use, according to this religious doctrine of the Middle Ages.

The fern called *Lunaria*, or moonwort, was held in great estimation, from a peculiar crescent shape of the *pinnæ* of its fronds, as a cure for lunacy and all diseases of a periodical character, especially for intermittent fevers. This crescent shape won it some astrological repute; and in order to preserve its virtues, it was to be gathered with a sacred observance of days. The moonwort was collected at the time of the full moon, and by the light of it, or its powers would be of no avail. Astrology was intimately blended with the practice of medicine in the Middle Ages, no less than theology, and many an herb was supposed to derive its healing powers from some tutelary planet. The most of the herbs in use by the ancient simplers were mere cordials. There were others of an entirely inert character that became famous from certain marvellous powers attributed to them by astrology. One of the most remarkable of these was the blue vervain, a conspicuous plant in fallow grounds and by-ways, flowering in August. So great was the reputation of this plant as a cure, that it bore the name of " simpler's joy," though now excluded as worthless from all standard pharmacopœias. The vervain was tied with a yard of satin ribbon around the neck, where it was to remain until the patient was cured. It was to be gathered at the rising of the dog-star, when neither the sun nor the moon shone, and with the left hand only. When thus collected it would vanquish fevers and other distempers, was an antidote to the bite of serpents, and a charm to conciliate friends after estrangement.

The healing virtues of many other herbs were ascribed to the planet under whose ascendency they were to be collected, and not to any intrinsic properties belonging to them. It was this belief in planetary influences that gave rise to the custom among physicians of prefixing to their recipes a symbol of the planet under whose light the ingredients were to be collected. A mistake in attending to the planetary hour would render these substances entirely inert. This fact may account for the vast number of inert remedies which have been popular in all ages. There was hardly a plant in medicinal use that was not believed to be under the auspices of some planet, and which must be gathered in strict accordance with the prescriptions of medical astrology.

In medical history nothing is more remarkable than the pertinacity with which mankind, through hundreds of ages, will cling to a supposed remedy, after it has been repeatedly tried and condemned as worthless by physicians. Men hug their medical notions in as close an embrace as the doctrines of their religious faith, and exercise their reason in regard to the one no more than in regard to the other. Indeed, the ancient union of prophet and physician in one profession caused medicine and religion to be intimately associated in the minds of the people. Hence the sanctity of an herb, caused by its consecration in certain religious ceremonies, was often considered better proof of its efficacy in the cure of diseases than any practical experience of its virtues. The remedial reputation of precious stones had a religious origin. They were supposed, on account of their purity and splendor, to be the residence of good spirits, and consequently useful as amulets to expel disease. These follies of human reason have not been wholly confined to the ignorant. The celebrated John Wesley, being worn down by excessive apostolic labors, visits the country, and after a few months' rustication is greatly relieved. He records this fact in his journal as the triumph of "sulphur and supplication" over his infirmities, and attributes his cure to daily prayers and a plaster of egg and brimstone, rather than to Dr. Fothergill's prescription of "country air, rest, milk diet, and horse exercise."

I am a believer in medicines and in medical science; and though quackery is a fated appendage of the healing art, as swindling and counterfeiting are the inevitable accompaniments of trade, and though it continues to cause great destruction of life, the loss of life would be still greater if medicines were entirely unknown and unemployed. But, as if intended as a safeguard to the dangerous arts of quacks, Providence has benevolently supplied the fields with thousands of innocuous herbs, and mercifully endowed mankind with faith in their remedial power, that they may amuse themselves, when sick, with harmless decoctions containing the semblance of physic in the guise of a cordial beverage. Many an honest person who was too ignorant to believe in medicine as a science — considering it but a supernatural gift bestowed exclusively upon the uneducated — has been saved from the malpractice of some charlatan by his faith in whiteweed and marigold, or in some equally harmless herb gathered at the rising of Sirius or under the waning light of the moon.

But there was no charlatanry among these charitable dames who brought balm to the sick, and dispensed their healing gifts without price. Some jealousy would occasionally arise between them and the learned faculty, from their interference in each other's jurisdiction; but they were seldom placed in direct antagonism. The balm, the mint, and the sage, brought to the patient by the considerate nurse, were often favorable accompaniments to the medicines presented by the physician. The simplers made the study of plants more of a utilitarian exercise than our present students, who admire flowers as beautiful objects, and study them as connected with taste and poetry. The modern student learns their technical characters, and examines their different parts as aids to the understanding of science. He pays but little regard to their medical virtues, which in most cases are but a part of the romance of their history. The experiments made and repeated, for the purpose of ascertaining the virtues of plants, thousands of times during several centuries, have enlightened the physician concerning their qualities, which are now very well understood. The simplers, however, supposed almost every plant to possess some quality designed for the sanitary welfare of the human race. Some old legend was associated with one, and some holy tradition with another, each pointing to the medical and magical virtues attributed to the plant and to certain benefits to be derived from it.

The herbalists among the early emigrants of Great Britain must have been greatly bewildered, when they went out into our American forests to seek the wild plants of their own native isle, and occasional unhappy accidents arose from false identification. When they discovered a plant that resembled any well-known English herb, they speedily declared the identity of the two, founding their judgment chiefly on the sensible qualities of the plants. It was by experiments of this class of botanists that the virtues of many of our indigenous herbs were determined. Not a few of our plants, however, owe their medical reputation to Indian traditions.

Among the recollections of my early life is that of the annual appearance of the herb-women, — vestiges of the ancient class of simplers, — who earned a livelihood, in part, by gathering and carrying to market herbs, roots, and flowers, to be used chiefly in the preparation of "diet-drink," a kind of small-beer, of which the bitter and aromatic herbs were the principal ingredients. In these packages were strips of white-pine bark, which in its dried state gives out the flavor of nutmegs, — slightly bitter and fragrant. The pitch-pine was also plundered of its recent shoots, before they were hardened into wood, and tied up with sweet-fern and the spicy leaves of the bayberry, and the root of sassafras. The umbelled pyrola, or rheumatism-weed, — a plant that bears several whorls of bright evergreen leaves, surmounted with an umbel of beautiful nodding flowers of purple and white, — also the yarrow and the roots of the yellow dock, were favorite ingredients, combined with the aromatic leaves of the checkerberry and St. John's wort. These careful dames, in the latter part of summer, employed themselves in collecting cordial herbs for winter's needs.

The herbs formerly gathered by the simplers are now cultivated in gardens devoted to this special purpose, belonging chiefly to the Shakers. All the romance attending the occupation is destroyed by this change. The herbs are now pressed into cakes and sold in the apothecary's shops.

I have never opened a package in which the slender, cordlike roots of the *Aralia nudicaulis* were wanting. The roots of the aralia closely resemble those of the true sarsaparilla, not only in their cordlike shape, but in their entire want of any medical virtue. It is remarkable that this entirely inert and tasteless root should be the only ingredient that is never omitted, and proves that any plant in use among popular remedies maintains its repute in proportion as it is destitute of medical properties of any kind. The same habits prevail among the semicivilized nations. The ginseng, for example, which is as inert as so much white paper, is regarded in China as a medicine that will cure all diseases. Tons of the roots of this plant are annually imported into that country. The ginseng is the popular panacea among the Celestials, and is held by them in the same estimation as sarsaparilla by the Americans. People will sometimes take efficacious remedies, when prescribed by their physicians; but no substance is mentioned in history which has acquired and maintained general popularity for any number of years, if it possessed any medical virtue at all. All curative drugs are unsafe, and if combined in a popular nostrum, soon excite mistrust, on account of accidents that happen from its maladministration. Many a patient, however, has been cured by mercury disguised by his physician in a preparation of sarsaparilla, without suspecting the cause of his cure.

A love of the marvellous also increases the popular faith in inert remedies. This innate propensity of the human mind formerly obtained gratification in mythological and magical superstitions. At present it finds more delight in mere abstractions that take no definite shape.

In the early ages the supposed marvellous effects of nihility were attributed to some planet, deity, or saint. Now they are equally credited, but referred abstractly to some hidden and mysterious power of nature. All the laws of nature are inexplicable; but nothing satisfies the general craving for the wonderful, unless it be impossible. It is not considered marvellous that a few grains of a poisonous substance should cause death, or that a smaller quantity of it should cure disease; but if it should be affirmed that an infinitesimal quantity of the juice of a plant whose juices can be swallowed by the pint without any effects upon the system, will cure disease, the assertion gratifies the popular appetite for the marvellous, and is believed.

It must be confessed that these old superstitions have spread the charm of romance over a great part of the vegetable kingdom. From these poetic illusions originated the ancient floral games and the use of plants in the ceremonies of religion, which is the great fountain of pure romance. The supernatural dangers that seemed to attend botanizing excursions of old enveloped all the wood in the charm of mystery. The mandrake was a plant whose destruction would be a forewarning of death to the person who should injure it. But as the mandrake was believed to possess some excellent properties for purifying the blood, which were indicated by its red sap, it was very desirable to be obtained as a medicine. An expedient was therefore adopted by the people to obtain possession of the plant, without implicating themselves. Its roots were fastened by a cord to some animal, usually a dog, who was compelled by whipping to pull them up from the earth. The dog was afterwards supposed to die, as a punishment for his involuntary act.

In these days we admire the peony as a splendid flower, and cultivate it in our gardens for its beauty. But the ancients imputed supernatural virtues to its roots; and as no medical property could be discovered in them, they were naturally supposed to be intended for a charm. Dr. Darwin writes, that even in his time bits of the dried roots of the peony were rubbed smooth and tied round the necks of children, to hasten the growth of their teeth. They were sold at the shops under the name of "anodyne necklaces." An ancient physician highly commends this necklace of the peony-root for the cure of epilepsy.

In the days of the Pythian oracles, when the priestess who delivered them was made drunk with an infusion of laurel-leaves before she prophesied, the sacred regard for the laurel in the popular mind must have equalled the reverence of the modern devotee for the shrine of the Virgin. The use of this decoction in the temple of Apollo, who was the god of music, poetry, and the arts, probably gave the laurel-tree its reputation as a crown for men of genius, and still later as a general crown of honor. The laurel, which is a dangerous narcotic, was never much employed as a medical remedy; and when it ceased to be used in the temples for purposes of divination, it was adopted as an evergreen for the brows of poets and heroes. But the age of romance has departed with the age of mythology, and the reverence that now attaches to these ancient superstitions is but the lingering twilight of a beauty that has passed away forever.

## ANGLING.

I HAVE often thought that the practice of angling was so intimately connected with the prospect of green fields and the smell of fresh meadows, that the general fondness for the sport originated in a great measure in our love of nature. I am so far, therefore, from considering the angler a model of patience, as Dr. Franklin regarded him, that I would rather look upon him as a sort of indolent devotee of nature, who prefers the voluptuous quiet of this sedentary sport to the more active habits of the gunner, the botanist, or the geologist. There are individuals, undoubtedly, who delight in angling from the love of the sport itself. Such are our inveterate fishers around the wharves and harbors, and who are generally better acquainted with the respective flavors of the different species of the finny tribe than with fishes as subjects of natural history. But the majority of anglers will be found to be genuine lovers of nature, and, like old Izaak Walton, as familiar with the plants that are growing at their feet, as with the little shining inhabitants of the lake and stream.

I am not of that sect of sentimentalists who would condemn angling on account of its cruelty. The pangs suffered by a little fish while expiring on the green bank are but momentary, and probably not to be compared with those of a bird when first taken from his native haunts and shut up in a cage. Fishes are but feebly endowed with the sense of feeling or touch, and have a brain so small as hardly to afford them a very definite consciousness. They have the senses of sight, of hearing, of smell, and of taste; for without these they could not provide for their own wants. They possess a very low form of intelligence and sensibility, and may be severely cut without showing signs of feeling. If we wound a bird, he may lead a life of pain and misery for many weeks. He is a creature of warm blood, endowed with intelligence and a capacity for grief. He is regarded as the companion and benefactor of man, and as having certain inalienable rights, — such as the enjoyment of life and liberty, and the means of obtaining a livelihood. But fishes, the voracious devourers of their own young, which they cannot recognize and do not protect, are plainly incapable of mental suffering, and may be taken in unlimited quantities without danger of causing an inconvenient scarcity. Hence, though all living creatures are more or less endowed with a power of feeling pleasure and pain, and have a certain right to the enjoyment of life, I regard the destruction of a fish in the same light as the killing of a fly or the trampling on a worm. I would not needlessly destroy an insect or set foot upon a worm; but I believe the united sufferings of a thousand fishes in the agonies of death would not equal the pangs suffered by one little child with a burnt finger.

There is no other sport so well adapted to the habits of a thoughtful man as that of angling, leading him out at noonday, under the shade of trees, or in the evening by the glassy stream, on whose mirrored surface he may view the surrounding hills and woods, while watching for the dimpling movements of the water that indicate the nibbling of the fishes. There can be no more delightful recreation in serene summer weather, when the heat of the atmosphere will not permit one to engage in more active toil or amusement. And there is no end to the pleasing fancies in which one may indulge the mind, while listening to the varied notes of the birds, that always frequent the borders of streams and lakes, or watching the motions of some little animal that will occasionally peep out upon one while occupied in his quiet amusement.

When we are seeking pleasure, it is not necessarily the prominent object of pursuit that is the source of the principal enjoyment we experience. Our object may be an errand of business in itself disagreeable, and our pleasures may spring from our adventures and observations during the time occupied in the performance of the errand. A walk is seldom interesting, however pleasant the scenery and other objects on the road, if we are sauntering without any particular aim. But if we have gone out to accomplish a certain purpose, which is of sufficient importance to keep up our resolution to proceed, many a scene on the road may be productive of a high degree of pleasure. Thus it seems to me that in angling the pleasure of the pursuit is, in almost all cases, derived from collateral circumstances, though the latter would be nothing without the purpose before us of taking our finny game.

The pleasure of angling consists in having some agreeable purpose to occupy the mind while indulging in the voluptuous sensations that attend us when surrounded by the charming accompaniments of green fields, fragrant woods, and pleasant prospects. To sit beside a stream for half a day, under the spreading branches of an oak, would be but a dull amusement for the most enthusiastic lover of nature, if he had no purpose in view except to enjoy the mere sensations derived from surrounding objects. But let him throw a hook and bait into the stream, with the intention of taking a few fishes to grace his table; and however insignificant their value, it is sufficient to furnish a motive for watching a float for many hours. The expectations which are thus aroused, and the agreeable exercise of the attention and the ingenuity, with the additional pleasure derived from the varied scenery, the fresh odors of vegetation, and the many cheerful sounds from animated nature, unite in rendering it one of the most interesting of employments.

Though I have never been a skilful or inveterate angler, the review of my angling excursions always brings before me some of my most pleasing recollections. The stillness of the occupation prepares the mind to receive impressions from adjacent scenes with singular vividness. The sight of the little fishes, as they are darting about among the long stems of the water-lilies, is then as pleasant to us as to a child. We watch every minute object with close attention, though it be but the little water-beetles as they whirl about on the surface near the shore, or the minute blossoms of the potamogeton that lift up their heads above the glassy wave. The lighting of a butterfly on the blue spikes of the pickerel-weed, or the humming of a dragon-fly, as he pursues his microscopic prey among the tall sedges and pond-weeds, never fails to attract our notice while engaged in our day-dreaming occupation.

While watching the float as it sails gently about with the wind, occasionally dimpling the surface of the water, we do not confine our attention to this alone. Not a bubble on the glossy sheen of the lake or the flitting shadow of a cloud as it passes over the sky escapes our notice. Everything that moves,—everything that can be seen or heard excites our curiosity as in the still darkness of night. When the fishes are inactive, as they often are during the heat of the day, we have little to do except to watch and observe the scenes and objects around us. At such times our attention is frequently attracted to something that hitherto might have been unobserved; and the squirrel that sits watching us from the bough of a neighboring tree, the little bird that is busy weaving stems, at no great distance, into the fork of a hazel-bush, and the sober cattle that have waded up to their knees into the shallow water, are all observed and studied with delight.

But the amusement of angling is not associated with sedentary observations alone; it is also connected with many interesting excursions in quest of more lucky fishing-ground. How often has it led us into delightful explorations of the woody boundaries of ponds, carrying us into seemingly impenetrable thickets, and causing the sudden discovery of some beautiful or curious plant, hitherto unknown to us, or introduced us to some new and rare species of bird or quadruped. It was on one of these rambles, by its musical and melancholy cooings, that I first discovered the wicker-nest of the turtledove, with its solitary egg, in the branches of a slender white-pine. On one of these occasions, also, I encountered for the first time the drooping fragrant flower of the Linnæa borealis, that exquisite production of northern climes, which is aptly named for the great Swedish botanist.

But the exercise alone, with the continual excitement of the curiosity, is sufficient to give interest to these excursions. Now we are led into green wood-paths, through fragrant bushes, some laden with flowers and others with fruit; now half bewildered by their intricacies, and then suddenly stumbling into a romantic view of the water and the surrounding scenery. Soon we pass into a deep dell, forming the bed of a stream, which has given rise to a multitude of rare and curious plants, and rouse the variegated summer duck from a solitary pool, embosomed in the thicket; finally, having arrived at an open pasture, a flock of sheep, startled at our approach, scamper off with resounding feet to a distant elevation. Then do we think with peculiar delight upon the pleasures of rural life, and regret that necessity which is ever leading us away from the abodes of peace and happiness. After performing a tour around the pond, we return perhaps to our original fishing-ground, pleased with the simple adventures we have encountered, and prepared to commence anew our patient toil.

As the decline of day begins to be apparent, the fishes are more active in their nibbling, and there is a more general stir among all the creatures of the field and wood. The thrushes are more musical in the neighboring thicket, and the yellow-throat comes within a few yards of us, and sings upon the branch of an alder-bush, as if he was pleased with our company. The frogs begin to be more loquacious, and our attention is attracted by different objects from those we observed at noonday or in the morning. A tortoise now and then protrudes its beak and eyes above the smooth sheen of the water, a little fish leaps out and makes a sudden plash, or a solitary snipe, with twittering notes, pursues its graceful flight along the shore.

At this time our luck as fishermen is usually the most propitious. The fishes that seem averse to the warm rays of the sun come out of deep water, as day declines, and look out for their prey, and are more active in nibbling the bait. After this time, in the space of half an hour, we often take fishes enough to make amends for any previous bad luck. Presently the float grows dim to the sight, the dew is perceptible on the grass, and the evening star, as it shines through the semicircle of light that surrounds the place where the sun went down, reminds us of home.

We prepare for our return, and for a change of scene and rest from our weariness; and home is never so delightful as it seems after one of these excursions. There is a luxury in our rest from toil which has been wearying but not exhausting; and the pleasures of social intercourse with our domestic circle are also greatly enhanced by a half-day's solitude. We partake of the bounties of our own table with a zest that proves it to be the design of Nature that man should toil for his subsistence if he means to enjoy the good things of her bounty. Thus terminates an amusement that brings us nearer to nature while we are engaged in it, that leads to pleasant observations and tranquil musings, while it prepares the mind to feel a renewed pleasure, when, wearied but not exhausted, we seek rest in the bosom of our family.

# AUGUST.

The plains and uplands are green with a second growth of vegetation, and nature is rapidly repairing the devastation committed by the scythe of the mower. But the work of the haymaker is not completed. He is still swinging his scythe among the tall sedge-grasses in the lowlands; and the ill-fated flowers of August may be seen lying upon the greensward among the prostrate herbage. The work of the reapers is also begun, and the sheaves of wheat and rye display their wavy rows to gladden and bless the husbandman. Flocks of quails, reared since the opening of the spring flowers, are diligent among the fields, after the reapers have left their tasks. They may be seen slyly and silently creeping along the ground, and now and then lifting their timid heads as if jealous of our approach. The loud whistling of the guardian of the flock, perched at a short distance upon a wall, may also be heard, and as we saunter carelessly along the field-path, a brood of partridges, rising suddenly almost from under our feet, will often astound our ears with their loud whirring flight.

Since the fading of the roses, the birds have generally become silent, as if the presence of these flowers were necessary to inspire them with song. They have grown timid and have forsaken their usual habits, no longer warbling at the season's feast or rejoicing in the heyday of love. They fly no longer in pairs, but assemble in flocks, which may be seen rising and settling over different parts of the landscape. Some species are irregularly scattered, while others gather into multitudinous flocks, and seem to be enjoying a long holiday of festivities, while preparing to leave their native fields. Their songs, lasting only during the period of love, are discontinued since it is past, and their young are no longer awaiting their care. On every new excursion into the fields I perceive the sudden absence of some important woodland melodist. During the interval between midsummer and early autumn one voice after another drops away, until the little song-sparrow is left again to warble alone in the fields and gardens, where he sang the earliest hymn of rejoicing over the departure of winter.

Since the birds have become silent, they have lost their pleasant familiarity with man, and have acquired an unwonted shyness. The warblers that were wont to sing on the boughs just over our heads, or at a short distance from our path, now keep at a timid distance, chirping with a complaining voice, and flee at our approach, before we are near enough to observe their altered plumage. The plovers have come forth from the places where they reared their young and congregate in large flocks upon the marshes; and as we stroll along the sea-shore, we are often agreeably startled by the sudden twittering flight of these graceful birds, aroused from their haunts by our unexpected intrusion.

It is now almost impossible for the rambler to penetrate some of his old accustomed paths in the lowlands, so thickly are they interwoven with vines and trailing herbs. Several species of cleavers with their slender prickly branches form a close network among the ferns and rushes; and the smilax and blackberry vines weave an almost impenetrable thicket in our ancient pathway. The fences are festooned with the blue flowers of the woody nightshade and the more graceful· plants of the glycine are twining among the faded flowers of the elder and viburnum. The lowlands were never more delightful than at the present time, affording many a pleasant arbor beneath the shrubbery, where the waters have dried away and left the greensward as sweetly scented as a bower of honeysuckles. In these places are we tempted to linger for refreshment on summer noondays, — bowers where it is delightful to repose beneath the shade of slender birches whose tremulous foliage seems to whisper to us some pleasant messages of peace. All around us the convolvulus has trailed its delicate vines, and hung out its pink and striped bell-flowers; and the clematis has formed an umbrageous trellis-work over the tops of the trees. Its white clustering blossoms spread themselves out in triumph above the clambering grape-vines, forming deep shades which the sun cannot penetrate, overhanging and overarching the green paths that lead through the lowland thickets.

When the pale orchis of the meads is dead, and the red lily stands divested of its crown; when the arethusa no longer bends its head over the stream, and the later violets are weeping incense over the faded remnants of their lovely tribe, then I know that the glory of summer has departed, and I look not until the coming of the asters and the goldenrods to see the fields again robed in beauty. The meeker flowers have perished since the singing-birds have discontinued their songs, and the last rose of summer may be seen in solitary and melancholy beauty, — the lively emblem of the sure decline of all the beautiful objects of this life, the lovely symbol of beauty's frailty and its transientness. When the last rose is gone, I look around with sadness upon its late familiar haunts; I feel that summer's beauty now is past, and sad mementos rise where'er I tread.

It is my delight to seek these last-born of the roses, and to my sight they are more beautiful than any that preceded them, as if Nature, like a partial mother, had lavished her best gifts upon these her youngest children. The bushes that support them are overtopped by other plants, that seem to feel an envious delight in concealing them from observation, but they cannot blot them from our memory, nor be admired as we admire them. The clethra with its white odoriferous flowers, and the buttonbush with its elegant globular heads, strive vainly to equal them in fragrance or beauty. The proud and scornful thistle rears its head close by their side, and seems to mock at the fragility of these lovely flowers; but the wild briar, though its roses have faded, still gives out its undying perfume, as if the essence of the withered flowers lingered about their former leafy habitation, like spirits about the places they loved in their lifetime.

In the latter part of the month we begin to mark the approaching footsteps of autumn. Twilight is chill, and we perceive the greater length of the nights and evening's earlier dew. The morning sun is later in the heavens, and sooner tints the fleecy clouds of evening. The bright verdure of the trees has faded to a more dusky green; and here and there in different parts of the woods may be found a sere and yellow leaf, like the white hairs that are interspersed among the dark-brown tresses of manhood, that indicate the sure advance of hoary years. The fields of ripe and yellow grain gleam through the open places in the woods, making a pleasant contrast with their greenness, displaying in the same instant the signs of a cheerful harvest and the melancholy decay of vegetation. The swallows assemble their little hosts upon the roofs and fences, preparing for their annual migration, and all things announce the speedy decline of summer.

Already do I hear at nightfall the chirping of the cicadas, whose notes are at the same time the harvest hymn of nature and a dirge over the departure of flowers.

When the evenings are perceptibly lengthened and the air partakes of the exhilarating freshness of autumn, these happy insects commence their anthems of gladness; and their monotonous but agreeable melody is in sweet unison with the general serenity of nature. These voices come from myriads of cheerful hearts, but there is a plaintiveness in their modulation that calls up the memory of the past and turns our thoughts inwardly upon almost forgotten joys and sorrows. How different are our emotions from those awakened by the notes of the piping frogs that hail the opening of spring! All these sounds, though not designed particularly for our benefit, are adapted by nature to harmonize agreeably with our feelings, and there is a soothing and lulling influence in the song of the cicadas that softens into tranquillity the melancholy it inspires and tempers all our sadness into pleasure.

We no longer perceive that peculiar charm of spring vegetation, that comes from the health and freshness of every growing thing; and we associate the flowers of August with the dry, withered, and dying plants that everywhere surround them. In June everything in the aspect of nature is harmonious; all is greenness and gladness, and nothing appears in company with the flowers to disfigure their charms or to affect the sight with displeasure. But August presents a motley spectacle of rank and inelegant weeds, that overshadow the flowers; and the beauty of the fields is often hidden by the withered vegetation of the last month. This appearance, however, is chiefly obvious in those places which have been disturbed by cultivation. In the wilds Nature always preserves the harmony of her seasons. Every herb and flower appears at proper time; and when one species has attained maturity it gives place to its rightful successors without any confusion, all rising and declining like the heavenly hosts of night, and clothing the face of the landscape in perpetual bloom and verdure. Seldom do we behold a parterre that equals in beauty those half-wild spots where, after a partial clearing of the forest, Nature has been left to herself a sufficient time to recover from the effects of art and to rear those plants which are best fitted for the soil and the season.

Let the lover of flowers and landscapes who would learn to gather round his dwelling all those rural beauties that will meet and blend in harmony receive his lesson from Nature in her own wilds. Let him look upon her countenance before it has been disfigured by a barbarous art, to acquire his ideas of beauty and propriety, and he will never mar her features by adding gems that do not harmonize with their native expression, plucked from the bosom of a foreign clime. Then, although he may not sit under the shade of the palm or the myrtle, or roam among sweet-scented orange-groves, in the climate of northern fruits and northern flowers, he needs no foreign trees or shrubbery to decorate his grounds or to adapt them to his pleasures. In a forest of his own native pines he may find an arbor in summer and a shelter in winter as odoriferous as a grove of cinnamon and myrtle; and the fruits of his own orchards will yield him a repast more savory than the products of the Indies.

# FORAGING HABITS OF BIRDS.

THE different habits of foraging that distinguish the several tribes and species of birds deserve attention as indicating a similar difference in the character of their aliment. Birds, for example, that take their food chiefly from the surface of the ground forage in a different manner from others that collect it from under the surface. Swallows catch all their food while on the wing, and give proof by this habit that they take only winged insects; but their manners differ essentially from those of the fly-catchers, that do not take their prey on the wing, but seize it as it passes by their perch. Robins and blackbirds gather their fare entirely from the ground, but their ways while seeking it differ exceedingly. Their respective habits of foraging are adapted to the successful pursuit of the worms and insects that constitute their principal diet. Though both species are consumers of all kinds of insects, they have their preferences, which are the chief objects of their pursuit. It is necessary to study their different habits of foraging to understand the principle which I have endeavored to inculcate, *that each species performs certain services in the economy of nature, which cannot be so well accomplished by any other species;* and that it is necessary for this end to preserve all in such proportions as would spontaneously exist if the whole feathered race were unmolested and left to their own natural chances of living and multiplying.

The sylvians are the most interesting of foragers among the smaller birds, and are remarkable for their diligence in hunting their prey. They have a peculiar way of examining the foliage and blossoms rather than the surface of the branches, and their motions are very conspicuous upon the outer parts of the trees near the extremity of the spray. The golden robin hunts his prey like the sylvians, though he is not one of them, and his motions are more rapid and energetic than theirs.

The wren, the creeper, and the chickadee seek their food while creeping round the branches, and take less of it from the foliage than the sylvians or the flycatchers. They seldom pause in their circuitous course, proceeding usually from the junctions of the branches to their extremities, hopping from spray to spray, and then passing to another tree. The sylvians appear to examine the leaves and blossoms, while the creepers and tomtits examine the bark of the tree. Hence the former do not prolong their stay with us after the fall of the leaf, while the other species are seen after the trees are entirely denuded. We may infer, therefore, that the sylvians feed chiefly upon beetles and other winged insects that devour the leaves of trees, while the creepers and tomtits take more insects in embryo, which during autumn and winter are half concealed in the bark of trees.

The habits of the flycatchers differ from those of any of the species above named. Let us take the pewee as an example. He sits on a twig almost without motion, but with a frequent sideling of the head, indicating his watchfulness. He does not seem so diligent as the sylvians, because he waits for his prey to come to him, and seeks for it only by carefully awaiting its approach. That he is not idle is shown by his frequent flitting out in an irregular circuit, and immediately returning to his perch with a captured insect. These salient flights are very numerous, and he often turns a somerset in the act of capturing his prey. He seldom misses his aim, and probably collects from ten to fifteen insects of an appreciable size every minute. As he lives entirely upon them, and in summer gathers them for his offspring, this is no extravagant estimate.

The pewee, however, does not catch all his prey while it is flying, but he is usually on the wing when he takes it. If he finds a moth or a beetle upon a leaf or a branch, he seizes it while he is poised in the air. A sylvian would creep along the branch, and when near enough extend his neck forward to take it. The vireos, forming an intermediate genus between the sylvians and the true flycatchers, partake of the habits of each. Some of them are remarkable for a sort of intermittent singing while hunting for their food. The preacher, indeed, seems to make warbling his principal employment. He is never, apparently, very diligent or earnest, and often stops during his desultory exhortations, to seize a passing insect, and then resumes his song.

Woodpeckers reside chiefly in the forest, of which they are the natural guardians; and as the food of their choice is nearly as abundant in winter as in summer, they are not generally migratory. Hence the operations of these birds are incessant throughout the year. As their food is not anywhere very abundant, like that of some of the granivorous birds, woodpeckers never forage in flocks. The more they scatter themselves the better their fare. The woodpeckers bear the same relation to other birds that take their food from trees, as snipes and woodcocks bear to thrushes and quails. They bore into the wood as the snipe bores into the earth, while thrushes and quails seek the insects that crawl on the surface of the ground.

There are several families of birds that take only a small part of their food from trees, and the remainder from the soil or the greensward. Such are all the gallinaceous kinds, larks, blackbirds, and thrushes. It has been said that the skylark was never known to perch upon a tree. These families are the guardians of the soil. The thrushes do not refuse an insect or a grub that is crawling upon a tree, but they forage chiefly upon the surface of the ground. In the feeding habits of the thrushes, their apparent want of diligence attracts frequent attention; but this appearance is delusive. The common robin will exemplify their usual manner, though he carries it to an extreme. When he is hunting his food he is usually seen hopping in a listless manner about the field. Sometimes a dozen robins or more may occupy one enclosure, but they are always widely separated. Observe one of them and you will see him standing still, with his bill inclined upward, and looking about him with seeming unconcern; soon he makes two or three hops, and then stands a few more seconds with his bill turned upward, apparently idle. Presently he darts suddenly a few yards from his standing-place, and may be seen pecking vigorously upon the ground. If you were near him you would see him pulling out a cutworm, seldom an earthworm, or devouring a nest of insects which are gathered in a cluster.

Blackbirds, though they also gather all their food from the ground, seem to be more industrious. Blackbirds of all species walk. They do not hop like the robin. They seldom hold up their heads, but march along with their bills turned downward, as if entirely devoted to their task. They never seem to be idle, except when a flock of them are making a garrulous noise upon a tree. If a blackbird looks upward, it is only by a sudden movement; he does not stop. After watching a blackbird and a robin ten minutes in the same field, any one would suppose that the blackbird had collected twice as much food as the robin during that time. But this is not true. The difference in their apparent industry is caused partly by the character of their food. The robin is entirely insectivorous, while the omnivorous blackbird hunts the soil for everything that is nutritious, and picks up small seeds that require a close examination of the ground.

The robin is probably endowed with a greater reach of sight than the blackbird, and while hopping about with his head erect, his vision comprehends a wider space. Many a time have I been astonished at the rapidity with which one of these idle robins would collect cutworms during a dry spell when they could not be very abundant, sometimes bringing two at a time in her bill and carrying them to her young. The robin not only watches for a sight of his prey, but also for the marks upon vegetation that denote the place of its concealment. He must possess an extraordinary share of this sagacious instinct; for the thousands of cutworms destroyed by him could not be discovered except by these indications and when they crawl out at twilight. The robin is therefore one of the earliest as well as the latest feeders among all our birds in the morning and evening.

The foraging habits of the different species of domestic poultry are worthy of remark as illustrating some of the differences observed in the manners of wild birds. Place a brood of ducks in a field during grasshopper-time, and they generally pursue one course, marching in a body over the field with great regularity. A brood of chickens, on the contrary, will scatter, occasionally reassembling, but never keeping close together, unless they are following a hen. Turkeys scatter themselves less than chickens, but do not equal ducks in the regularity of their movements. Pigeons settle down upon a field in a compact flock, and then radiate in all directions. They pursue no regular march, like ducks.

A very interesting class of foragers are those that feed in compact assemblages. This habit renders the snow-buntings exceedingly attractive. Their food is not distributed in separate morsels like that of robins and woodpeckers. It consists of the seeds of grasses and of composite plants, which are often scattered very evenly over a wide surface. When, therefore, a flock of fifty or more settle down in a field, each one fares as well as if he were alone, during the short time they remain on the spot. Insect-feeders find it for the most part profitable to scatter and keep separate, because their food is sparsely distributed. This is not true of the birds that frequent the salt-marshes that are overflowed by the tide. Their aliment consists of insects and worms which are evenly scattered and abundant. Hence sandpipers and some other species forage in flocks, though they live exclusively upon an animal diet.

The swallow tribes are the guardians of the atmosphere, that would otherwise swarm with fatal quantities of minute insects. Their foraging habits are observed by all, and are well known. Woodpeckers, creepers, and chickadees are the guardians of the timber of the forest; sylvians and flycatchers, of the foliage. Blackbirds, thrushes, crows, and larks are the protectors of the surface of the soil; snipes and woodcocks, of the soil under the surface. Each family has its respective duties to perform in the economy of nature; and man must beware lest he disturb this equilibrium by reducing the numbers of any species below the supply of insects which is afforded them.

It is curious to note the assiduity with which insects are hunted in all stages of their existence. In their larva state, those that lurk inside of the wood and bark are taken by woodpeckers, and those under the soil by snipes and woodcocks. Insects, when the larva has assumed the form of moths, beetles, and flies, are attacked by flycatchers and sylvians and other small birds that take their food by day, and by small owls and whippoorwills by night. It matters not in what stage of its existence the insect is destroyed; it is still demonstrable that these minute creatures cannot be kept in check unless they are attacked in all stages. Birds are their only effectual destroyers. Man cannot check their multiplication or their ravages by artificial means. He cannot even protect his garden. Their destructive and infinite multiplication can be prevented only by Nature's own agents, which she has created with this power. A million of ichneumons would not do the work of a dozen birds.

# BIRDS OF THE AIR.

ALL birds that take their food while on the wing, and seldom or not much in any other way, may be arbitrarily designated as Birds of the Air, whether their prey inhabit the air, like the insects taken by the Swallows and Flycatchers, or the cup of a flower, like those taken by the Humming-Bird. Of these the Swallows, including the Martin and the Swift, are the most conspicuous and most numerous in this part of the world. These birds have large wings, fly very swiftly, and without a great deal of apparent motion of their wings. It could hardly be explained on mechanical principles how they are able to pass through the air with such rapidity. While watching them on the wing, it seems as if they were never weary; but Daines Barrington says the Swallow makes frequent pauses for rest while engaged in the pursuit of insects.

### THE BARN-SWALLOW.

This is the species with which the inhabitants of New England are best acquainted. But they are every year becoming fewer, and this diminution of their numbers is attributed by Mr. S. P. Fowler to our modern tight barns. Though they often build under the eaves of houses and in sheds, they find in these places but limited accommodations, compared with the old-fashioned barns that were formerly scattered over the whole country. There are now hundreds only where thirty years ago there were

# CRITICAL
# ISSUES

thousands, all swarming with these lively birds, who built their nests on the horizontal beams that supported the barn roof. The birds left us when they were deprived of their tenements, while the Cliff-Swallow, that builds under the eaves of barns and houses and under projecting cliffs of rocks, has increased, feeding upon the larger quantity of insects consequent upon the absence of the Barn-Swallow.

This species is of a social habit; fond of building and breeding, as it were, in small communities. An old-fashioned barn has been known to contain as many as two dozen nests. They are constructed of materials similar to those of a Robin's nest; but the Swallow adds to the lining of grass a few feathers, which the Robin does not use. Dr. Brewer alludes to a custom among the Barn-Swallows of building "an extra platform against, but distinct from the nest itself, designed as a roosting-place for the parents, used by one during incubation at night or when not engaged in procuring food, and by both when the young are large enough to occupy the whole nest." The eggs of the Barn-Swallow are nearly white, with a fine sprinkling of purple. Two broods are reared in a season. When the bird appears to have a third brood I think it must have happened from the accidental destruction of the second brood of eggs.

### THE CLIFF-SWALLOW.

The Cliff-Swallow is the species that has apparently filled the vacancy made by the diminished numbers of the Barn-Swallow. It is a smaller bird and more whitish underneath. The nests of this species are placed under the eaves of houses, sometimes extending nearly across the whole side of a roof, resembling in some degree a long row of hornets' nests. The nest is of a roundish shape; the body of it is plastered to the wood; the entrance is the neck, slightly covered for protection from rain. They are made of clay and mud without intermixture of other substances. They are lined with grass and feathers.

This species was at the early settlement of the country so rare, in this part of the continent, that it escaped the notice of some of the earliest observers of the habits of our birds. It was not known even to Alexander Wilson. It seems to have been observed and described in Maine before it was well known in any of the other States. Dr. Brewer says of this species : " I first observed a large colony of them in Attleborough (Mass.) in 1842. Its size indicated the existence of these birds in that place for several years. The same year they also appeared in Boston, Hingham, and in other places in the neighborhood." The notes of this Swallow are not so agreeable as those of the Barn-Swallow and other species.

### THE WOOD-SWALLOW.

The White-bellied Swallow is known in the British Provinces by the name of " Wood-Swallow." This will be regarded a very appropriate designation, when we consider the continuance of the primitive habits of this bird of building in hollow trees. Samuels has seen great numbers of the nests of this species in the woods of Maine, near the northern lakes, built in hollow trees, some of them standing in water. In an area of about ten rods he counted fifty nests. He says this species is the most common of the Swallows in that region. The nests are formed entirely of grass and feathers without any mud, for which there is no necessity. The eggs are pure white.

This species has superseded the Purple Martin in many

parts of New England, as the Cliff-Swallow has superseded the Barn-Swallow. They are pretty generally distributed over the whole continent, though, notwithstanding the primitive habits that still adhere to a great part of their numbers, they are most numerous in cities and their suburbs, attracted probably by the vast multitudes of small flies, which are more abundant than in the woods. The Cliff-Swallow breeds as far as the Arctic Seas.

THE SAND-MARTIN.

This is not the least interesting of the family of Swallows. The swarming multitudes that often assemble in one vicinity, their constant motions while going in and out their holes in the sand-bank, and sailing about on rapid wing in quest of their microscopic prey, and their lively notes render them objects of frequent attention. Of all the Swallows the Sand-Martins afford the most amusement for small boys in the vicinity, who employ themselves in digging out their nests, which are sometimes less than two feet under the surface. The difficulty is in finding the exact spot where the excavation should be made. Large multitudes of them formerly assembled every year and made their holes in the high sand-bluffs that surround the Beverly coast. I have counted over fifty holes in one large and high bank.

"The work of preparation," says Dr. Brewer, "they perform with their closed bill, swaying the body round on the feet, beginning at the centre and working outwards. This long and often winding gallery gradually expands into a small spherical apartment, on the floor of which they form a rude nest of straw and feathers. The time occupied in making these excavations varies greatly with the nature of the soil, from four or five days to twice that number."

## THE PURPLE MARTIN.

It is seldom in these days we hear the sweet hilarious notes of the Purple Martin in Eastern Massachusetts. From some not very accountable cause the species have left many of their former habitations, and we are no longer pleasantly roused from our sleep by their sportive garrulity near our dwellings. The absence of these birds is a truly sorrowful bereavement. When I visit the places where I formerly heard them and note their absence, I feel as I do when strolling over some old familiar ground upon which every scene has been changed, where wood has become open space, old houses are removed and replaced by new, and strangers occupy the homes of the old inhabitants.

We no longer see any large assemblages of Purple Martins in Eastern Massachusetts; and in almost all parts of New England, where they were formerly the most common of our birds, their numbers are greatly diminished. Why, it may be asked, have they so generally left these parts, especially the vicinity of Boston? May it not be that the Wood-Swallows, which have multiplied in the same ratio as the Purple Martins have decreased, have been the cause of their disappearance? They breed in the boxes formerly used by the Martins, who, upon their later arrival, finding them preoccupied by the Wood-Swallow, and failing to obtain other accommodations, fly away to another vicinity. In a contest for a box the Purple Martin would be the victor, but would prefer seeking a habitation elsewhere to making an attempt to dislodge birds which had already built their nests there.

The Purple Martin is the largest of the American Swallows, with plumage of a bluish-black intermingled with purple and violet. In beauty it is not surpassed by any of the species. It seems to have no fear of man, who from immemorial time has protected it. The aboriginal inhabitants set hollowed gourds upon the trees to draw the Martins to their huts. And when the white man came, he provided them with a meeting-house, considering it a fitting structure for their musical congregations.

The Purple Martin utters a series of notes which are so varied and continued as to deserve to be called a song. This song has attracted less attention from those who have described the habits of our birds than it merits. In my early days I have listened for hours to the peculiar notes of the Purple Martin, in which a variety of chattering and chuckling is combined with a low guttural trill, resembling certain parts of the song of the Red-Thrush. The Martin, however, does not give himself up to song. His notes are heard chiefly while on the wing; but they are almost incessant. He is constantly in motion, and his song seems to me one of the most animated and cheerful sounds uttered by any American bird except the Bobolink.

The flight of the Purple Martin and his peculiar ways render him exceedingly interesting and amusing. Surpassed by no bird in swiftness, there is none that equals him in the beauty of his movements on the wing, uniting grace and vivacity in a remarkable degree. Often skimming the surface of ponds, or swiftly gliding along a public road a few feet from the ground, then soaring above the height of the lower clouds, he sails about with but little motion of the wings, till he is out of sight. These flights seem to be made for his own amusement; for it cannot be supposed that he finds the larger insects that constitute his prey at so great a height.

The boldness displayed by the Purple Martin in driving Hawks and Crows from his neighborhood accounts for the respect in which he was held by the Indians, who were great admirers of courage. " So well known," says Wilson, "is this to the lower birds and to the domestic poultry, that as soon as they hear the Martin's voice engaged in fight, all is alarm and consternation." The Martin is often victor in contests with the Kingbird, perhaps when one is tired of the contest another takes his place with fresh vigor, so that the Kingbird is finally driven away and conquered.

### THE CHIMNEY-SWALLOW.

The Chimney-Swallow attracts general attention on account of its practice of building its nest in the unused flue of a chimney. In village and town this family of birds are very abundant, some deserted chimney being always appropriated for the rearing of their young. It is remarkable that their desertion of their original breeding-places and their present selection of chimneys should be so universal. Though they are known at the present time to build, as formerly, in hollow trees, they do so only in forests very distant from town or village. It cannot be said that they are fond of the companionship of man. The small flies that constitute their food are probably more numerous in towns than in forests. Hence the birds for convenience resort to the chimney rather than the hollow tree, which is farther from their supplies of food.

The Chimney-Swallow is the smallest of our American species, and is partially nocturnal in its habits, being most active during morn and early twilight. Its nests are nicely woven with sticks, fastened to the chimney with a glutinous saliva. Says Samuels: "About sunset, great multitudes of these birds are out, and the numbers of insects they destroy must be immense. Everywhere they may be seen; away up in the blue sky, as far as the eye can reach, they are coursing in wide-extended circles, chasing each other in sport, and even caressing and feeding their mates while on the wing. A little lower they are speeding over the tops of trees, gleaning the insects that have just left the foliage; over the surface of the lake or river they fly so low, in the pursuit of aquatic insects that their wings often touch the water. Everywhere are they busy."

### THE KINGBIRD OR BEE-MARTIN.

The true Flycatchers take all their food while it is flying in the air, though they do not sail round, like a Swallow, to catch it. They are commonly seated quietly on their perch, and seize it by sallying out a few yards, and then returning. If we watch the ways either of the Kingbird or the Pewee, we shall observe this peculiar habit of all the Flycatchers. One of the most common of our birds, well known by his lively manners, his shrill notes, and twittering flight; always apparently idle, sitting on the branch of a tree as if he were a sentinel of the field, is the Kingbird. From this branch you may observe his frequent sallies when darting upon his prey. You may often see him pursuing a Hawk or a Crow, and annoying it by repeated attacks, always made in the rear of his victim. His usual custom is to rise a little above the object of his harassment, and then swoop down in such a manner that the bird cannot turn upon him. I have frequently seen him rise almost out of sight when engaged in such encounters. His victim constantly endeavors to rise above his pursuer, while the Kingbird by his activity as invariably balks him. I could never determine which of the two was the first to tire. But the Kingbird may probably be relieved by another of his species who may take his place. This pugnacious habit is said to continue only during the breeding-season.

It is amusing to watch his movements when flying. He sails rapidly along the air with but little motion of his outspread wings, save the vibrations of his extended feathers, all the time screaming with a sharp and rapid twitter. You observe this habit of the bird at short distances from the ground, when pursuing an insect. Upon seizing it he returns immediately to his post. He is watching all the while for the larger insects. He will not quit his perch, upon a fence, the branch of a tree, or a mullein-stalk, to catch small flies. He leaves all minute insects to the Swallows and small Flycatchers. The farmers complain of him as a bee-eater, whence the name of Bee-Martin which is often applied to him. Some observers say he discriminates between the different kinds of bees, selecting only the drones for his repast. But among the offences charged against him, he is never accused of stealing grain or fruit. Hence he is seldom molested, and enjoys great security compared with many other equally useful birds.

The Kingbird has not much beauty of plumage; but he is so neatly marked with black and white, with a bluish color above, and a white band at the extremity of his dark tail-feathers, and he displays his form and plumage so gracefully in his vibrating flights, that he cannot escape notice. The crest, containing a vermilion centre, is hardly discernible, save when the bird is excited, when it is slightly elevated. The Kingbird more frequently builds in an orchard than in a wood, an open cultivated place being more productive of those insects which afford him subsistence.

### THE PEWEE.

If we stroll at any hour of the day in summer and sit under a rustic bridge for coolness or shelter, while watching the stream and listening to its flow, we may hear the plaintive cry of the Pewee, a common but retiring bird, whose note is familiar to all. He seems to court solitude, though he has no apparent fear in the presence of man; and his singular note harmonizes with the gloominess of his retreat. He sits for the most part in the shade, catching his insect prey without any noise, but after seizing it, resuming his station. This movement is performed in the most graceful manner; and he often turns a somerset or appears to do so, if the insect at first evades his pursuit. All this is done in silence, for he is no singer. The only sound he utters beside his *lament* is an occasional clicking chirp. All the day, after short intervals, with a plaintive cadence he modulates the syllables *pe-wee*. As the male and the female can hardly be distinguished, I have not been able to determine whether this sound is uttered by both sexes or by the male only.

So plainly expressive of sadness is this remarkable note, that it is difficult to believe the little creature that utters it can be free from sorrow. Certainly he has no congeniality with the sprightly Bobolink. Why is it that two simple sounds in succession can produce an effect on the mind as intense as a solemn strain of artificial music and excite the imagination like the words of poesy? I never listen to the note of the Pewee without imagining that something is expressed by it that is beyond our ken; that it sounds in unison with some one of those infinite chords of intelligence and emotion, which in our dreamy moments bring us undefinable sensations of beauty and mystery and sorrow. Perhaps with the rest of his species, the Pewee represents the fragment of a superior race which, according to the metempsychosis, have fallen from their original high position among exalted beings; and this melancholy note is but the partial utterance of sorrow that still lingers in their breasts after the occasion of it is forgotten!

Though a retiring bird, the Pewee is very generally known on account of his remarkable note, which is heard often in our gardens as well as in his peculiar habitats. Like the Cliff-Swallow, he builds his nest under a sheltering roof or rock, and it is often fixed upon a beam or plank under a bridge. There are no prejudices in the community against this species. They are not destroyed on any occasion. By the most ordinary observer they cannot be suspected of doing mischief in the garden. I should remark in this place, that the Flycatchers and Swallows and a few other species that enjoy immunity in our land, though multiplied to infinity, would perform only those offices which are assigned them by nature. It is a vain hope that while employed in exterminating any species of small birds their places can be supplied and their services performed by other species which are allowed to multiply to excess. The Swallow and the Pewee, with all their multitudinous families, will not perform the work of the Robin or the Woodpecker, nor can all these together do the work of the Sylvians.

### WOOD-PEWEE.

We seldom ramble in a deep wood without hearing the feeble and plaintive note of the Wood-Pewee, — a bird that does not leave the forest, and is therefore less known than the larger species that builds under bridges and the eaves of old houses. The Wood-Pewee places its shallow nest upon some large branch of a tree without any protection above it, and it is chiefly concealed by the resemblance of its materials to the mosses and lichens on the bough. Its habits, except its attachment to the solitude of the wood, differ but little from those of the common Pewee. It seems likewise to have the same cheerful manners. The minor notes of the two Pewees serve, more than any others equally simple, to harmonize the anthem of Nature.

### THE HUMMING-BIRD.

The Humming-Birds, of which it is said there are more than four hundred species, are among the most exquisite of all animated beings. They unite the beauty and delicacy of a beautiful insect with the organization and intelligence of a creature of flesh and blood. Of all the feathered tribe, none will compare with them in the minuteness of their size. The splendor, variety, and changeableness of their hues are no less admirable than their diminutiveness. The colors of the rainbow do not surpass those of many of the species either in beauty or variety. A brilliant metallic lustre greatly enhances all this splendor. The variability of their hues, which is also observed in many other birds, is in the Humming-Birds almost unaccountable. Says Dr. Brewer: "The sides of the fibres of each feather are of a different color from the surface, and change as seen in a front or an oblique direction; and, while living, these birds by their movements can cause their feathers to change very suddenly to different hues. Thus the *Selasphorus rufus* can change in a twinkling the vivid fire color of its expanded throat to a light green; and the species known as the Mexican Star, changes from a light crimson to an equally brilliant blue."

Yet with all their beauty of color, what is most attractive about them is their flight. When a Humming-Bird is flying, so rapid are the motions of its wings that it seems like the body of a bird suspended in a circle of radiating sunbeams, or like one in the midst of a globe of down, like that which surrounds the receptacle of a ripened dandelion flower. When we watch the flight of a short-winged bird like the Quail, the radiations formed by the rapid motions of its wings make only a semicircle. In the Humming-Bird they form a complete circle of luminous rays. This flight, which resembles that of certain insects, is the more remarkable on account of the extraordinary length of its wings, which would lead us to infer that they would be incapable of such rapid motion by the muscular force of so small a body. The wings of those moths and beetles which have a similar movement bear no proportion to the length of the Humming-Bird's wing, compared with the size of the body of the insect and of the bird. It is the rapid vibration of the wings, producing a sound like the spinning of a top, that has given to this family of birds the name by which they are designated.

While hovering before a flower, this hum is plainly audible; but when the bird darts off to another place the tone produced by these vibrations is plainly raised to a higher key, as it spins like an arrow through the air. Dr. Brewer, alluding to the Swiss philosopher Saussure, says: " On the first visit of this naturalist to a savanna in the island of Jamaica, he noticed what he at first took to be a brilliant green insect, of rapid flight, approaching him by successive alternations of movements and pauses, and rapidly gliding among and over the network of interlacing shrubs. He was surprised by the extraordinary dexterity with which it avoided the movements of his net, and yet more astonished to find, when he had captured it, that he had taken a bird and not an insect."

The largest known Humming-Bird is about the size of the Chimney-Swallow; and so great is the disparity in the size of the different species, that when confined in a cage, and the perch "has been occupied by the great

Blue-throated Humming-Bird, the diminutive Mexican Star has settled on the long beak of the former, and remained perched on it some minutes without its offering to resist the insult." Some of the species are so small that if they flew by night they might be swallowed alive by one of the smaller Owls as easily as a beetle.

The Humming-Bird was formerly supposed to feed entirely on the nectar of the flowers it was seen so constantly to visit. It is now well ascertained that its chief subsistence is made up of small insects which it takes from the flower. But the ancient opinion was not entirely a fallacy, since a portion of the nectar of the flower is taken with the insects, and supplies to the Humming-Bird that kind of nourishment which the larger insectivorous birds derive from fruit. Dr. Brewer says "the young birds feed by putting their own bills down the throats of their parents, sucking probably a prepared sustenance of nectar and fragments of insects." The bird uses his tongue both for capturing insects and for sucking the drops of dew and nectarine juices contained in the flower.

Notwithstanding the small size of the whole tribe of Humming-Birds, they are notoriously the most courageous and combative birds in existence. Their sharp bills, their rapid flight, the electric quickness of their manœuvres, render them so dangerous that no bird whom parties of them choose to attack can escape unharmed.

I once discovered a nest of the Humming-Bird in my own garden, upon the horizontal bough of an old apple-tree. It was placed near the end of the bough, about five feet from the ground. It was built, as all writers have described other nests of Humming-Birds, of ferns and mosses, with lichens glued together, perhaps from being collected while they were damp. It contained two eggs about the size of a pea-bean.

# SWALLOWS: THEIR HIBERNATION.

THERE is not much that is interesting to be said of swallows, which are not singing-birds, and do not by their aerial flights attract attention, as if they were seen creeping on the branches of trees, and associated with their flowers. We watch with admiration their rapid movements through the air, their horizontal flight along the surface of some still water, and are charmed with their twittering when assembled round their nests. There was once a lively controversy in relation to the manner in which swallows pass the winter. The opinion of naturalists in Sweden and in the North of Europe, among whom we may name Linnæus and Kalm, was that swallows buried themselves in water under the freezing-line, or slept in the crevices of rocks. This theory has been discarded by modern naturalists, who have authentic accounts of flocks of swallows which have settled upon the masts and sails of ships when on their passage to or from the countries where they pass the winter. Still, the mystery is not cleared up.

White of Selborne mentions a week in March that was attended by very hot weather, when many species of insects came forth, and many house-swallows appeared. On the immediate succession of severe cold weather, the swallows disappeared and were seen no more until April. He mentions another instance recorded in his journal, of the reappearance of swallows after a month's absence, on the 4th of November, just for one day, which was remarkably warm, playing about at their leisure, as if they were near their place of retreat. On the same day, more than twenty house-martins appeared, which had retired without exception on the 7th day of the October previous. He adds that whenever the thermometer is above 50°, the bat flits out during any autumn or winter month. The author concludes that two whole species of swallows, or at least a large proportion of them in Great Britain, never leave the island, but remain torpid in some place of retreat; for he remarks, "We cannot suppose that, after a month's absence, house-martins can return from Southern regions, *to appear for one morning in November;* or that house-swallows should leave the districts of Africa, to enjoy in March the transient summer of a couple of days."

Daines Barrington testifies that he has in many instances known martins to reappear during warm days in different parts of the winter, but he is not sure that he has ever seen swallows at such times. He thinks, therefore, that martins conceal themselves in crevices of rocks, from which on a warm day they can emerge; but swallows, which are buried under water, cannot feel the influence of a short period of warm weather. The treatises on Ornithology written in the northern parts of Europe allude frequently, as if it were an established fact, to the submersion of swallows during the winter. Peter Brown, a Norwegian painter, informed Mr. Barrington that while he was at school near Sheen, he and his comrades constantly found swallows in numbers torpid under the ice that covered bays, and that they would revive if placed in a warm room. The author of a paper read before the Academy of Upsal mentions the submersion of swallows as a known fact in that part of the world. Among the superstitions associated with this belief, Pantoffidan relates that swallows before they sink under water sing the *Swallow Song*, as it is called, and which everybody knows.

A gentleman of science informed Mr. Barrington that when he was fourteen years of age, a pond belonging to his father, who was a vicar in Berkshire, was cleared out in February. While the workmen were clearing it, he picked up a cluster of three or four swallows that were caked in the mud, and they revived and flew about when carried to a warm room. Mr. Barrington records many similar facts, for which I have no space. In one instance swallows were taken out of a mass of solid ice, and were brought to life by the application of heat.

He thinks swallows only are ever submerged in water or mud, but that martins retire to fissures in rocks or to some lurking-places in the ground. He mentions a boatman who had seen thousands of martins in the crevices of a rock, and that they would revive when taken into a warm room. Kalm also relates, in his "Travels in America," that they have been found torpid in holes and clefts of rocks near Albany, New York. Mr. McKenzie, being at Lord Stafford's in Yorkshire, near the end of October, a conversation began about swallows crossing the seas. This the game-keeper disbelieved, and said he would carry any one to some neighboring coal-works, where he was sure of finding them at that time. Some of the servants attended him to the coal-pits, where several martins were found in a torpid state, but would show life when warmed.

Mr. Barrington concludes from all these facts that martins appear occasionally throughout the winter, when the weather is mild; but he had heard no well-attested cases of the reappearance of sand-martins during the winter; he cannot conjecture where they conceal themselves, but he is positive they do not winter in their holes. He expresses his belief in the impossibility of their making a journey across the seas to Africa, and doubts the few recorded instances of their alighting on the masts of vessels on their journeys of migration. If this theory of the migration of swallows be true, it must be true of those in the northern and southern parts of Asia. On the contrary, they hide themselves in the banks of the Ganges, during the three so-called winter months in that part of the world. Du Tertre mentions that the few swallows seen in the Caribbee Isles are only observed in summer, as in France. We are assured by Dr. Pallas, that not only are there swallows in Russia and Siberia, but that on the banks of the Wolga, latitude 57°, they disappeared about the fourth of August. These birds, according to the theory of migration, ought to have been passing to the more southern parts of Asia. Yet it has not been observed by any Asiatic traveller that they have the same species of swallow, or that they are seen in those parts during our winter.

As an objection to the theory of the torpidity of swallows as their mode of hibernation, it is asked where and when they moult, if not in regions south of Europe, as they do not moult before their disappearance. This is an objection that Mr. Barrington fails to answer. It is impossible, however, that their moulting can happen when submerged in water or torpid in some concealed resort. The functions of the animal economy would be unable to supply a new plumage while the system is in this state. I would suggest, if the theory of their torpidity were proved, that they may drop their feathers one by one, during all their active season of flight, as human hair is shed. Still, I cannot but think it more probable that swallows leave their northern habitats very early in the season, that they may arrive at their winter-quarters just before the season of moulting; and that the cause of their remaining undiscovered during their residence in the warm regions to which they resort is, that while moulting they live upon the ground in shelters of thicket, not being able to fly, and subsist upon a diet which they pick up from the ground.

But this does not explain the moulting of those swallows and martins, few or many, which have been proved to remain torpid in northern countries. Do these come out in the spring only to die, or do they perish in their winter retreats and never revive? If they are destined to perish here, why has Nature provided them with an instinct which answers no purpose whatever in their economy? If this submersion is only a method of suicide, why do they not perish immediately, instead of lingering along during the whole winter to die at the end of this season? And if they do not perish at this time, but awake and revive like bats and dormice, the most important question is, not where and when they moult, but why Nature has provided migration for a part of each swallow family, and a torpid sleep under water, and in crevices of rocks, for the remainder of the same families. I cannot but conclude that there is yet the greatest burden of proof remaining with those who maintain the theory of migration.

# THE FLOWERS OF AUTUMN.

THE student of Nature, who is accustomed to general observation, cannot fail to have noticed the different character of the flowers of spring, summer, and autumn. Each season, as well as climate, has a description of vegetation peculiar to itself; for as spring is not destitute of fruits, neither is autumn of flowers, though they have in general but little resemblance to one another. Those of spring, as I have already remarked, are delicate and herbaceous, pale in their tints, and fragrant in their odors. The summer flowers are larger, more brilliant in their colors, and not so highly perfumed as those of spring. Lastly, the flowers of autumn appear in unlimited profusion, neither so brilliant as the former, nor so delicate as the latter. They are produced on woody stalks, often in crowded clusters, and nearly destitute of fragrance. The differences in the general characteristics of the flowers of different seasons are an interesting theme of speculation; and they represent, somewhat imperfectly, the flowers of the different latitudes.

But there are certain species, appearing late in the autumn, that display the characters of the spring flowers, like the neottia, the purple gerardia, and the hedge hyssop.

The summer flowers are in their greatest splendor in the latter part of June. The greater number of those which commence their flowering in August are autumnal, and do not acquire their full maturity until September. The summer flowers are characterized by their large size and brilliant colors, and combine the two qualities of delicacy and splendor in a greater degree than those of any other season. Such are the different species of the beautiful orchis tribe, the cardinal-flower, the cymbidium, the arethusas, and some of the wild lilies. The majority of the flowering shrubs put out their blossoms in early summer, just after the blossoming of the fruit-trees. These diminish in number as the summer advances, and in autumn hardly one is to be found that is not loaded with seeds or fruit. The flowering plants of autumn, however, though not shrubs, are woody in their texture, and many are, in fact, a kind of annual shrubbery.

The summer flowers may be said to date their commencement with the elegant Canadian rhodora, and to end with the alder-leaved clethra, a flowering shrub very common in our swamps, bearing long slender spikes of white blossoms which have the odor of lilacs. During this interval, the most beautiful flowering shrubs of our climate unfold their blossoms. The rhodora is followed in succession by the honeysuckles, the kalmias, or false laurels, the azaleas, the viburnums, and many others not less important as ornaments of our native landscape. The flowering of the alder-leaved clethra marks the decline of summer. After this, the remainder of the month of August is a period rather barren of wild-flowers. The most of those which are peculiar to summer have faded, and the autumnal tribes are still ripening their buds. There seems to be a short suspension, during this month, of the efforts of Nature, while she is preparing to unfold the brilliant treasures of autumn.

The spring produces in the greatest abundance those flowers that affect a northern latitude. As the season advances we find more of those tribes which are peculiar to warm climates. The roses and rosaceous flowers usually appear in the early summer weeks, and the flowers of these genera are rare in tropical regions, being the denizens chiefly of temperate latitudes. The papilionaceous flowers, of which the greater numbers of species are found within the tropics, do not appear with us in profusion until the latter part of summer. The prevailing hues of the summer flowers are the different shades of scarlet, crimson, and purple, which grow paler as the days decrease in length and the temperature becomes cooler. Thus the bulbous arethusa, that flowers in June, is of a brilliant purple or crimson; while the adder's-tongue arethusa, that appears a month later, is of a pale lilac. Our native species of the brightest tints belong to summer. Such are the scarlet lobelia, the narrow-leaved kalmia, the red lily, and the swamp rose.

With August appears a kind of vegetation unlike any that has preceded it. The compound flowers, a very extensive tribe, begin to be conspicuous. These flowers are characteristic of vegetation in the autumn, the greater part of them coming to perfection during this season, beginning with a few species in the month of August. All these increase in beauty and variety until September arrives, bearing superb garlands of asters, sunflowers, and goldenrods, which, though exceeded in delicacy and brilliancy by the earlier flowers, are unsurpassed in splendor. The season of the autumnal flowers may be dated as commencing with the flowering of the trumpet-weed, or purple eupatorium. This is one of the most conspicuous plants in our wet meadows during the early part of September. It often grows perfectly straight to the height of six feet, in a favorable soil, bearing at regular distances around its cylindrical stem a whorl of leaves, which by their peculiar curvature give the plant a fancied resemblance to a trumpet. Soon after this appear the yellow gerardias, bringing along with them countless multitudes of asters, goldenrods, and autumnal dandelions, until the uplands are universally spangled with them, and gleam with a profusion of blossoms unwitnessed at any other season.

The asters are the most remarkable of the flowers of autumn, and are, in many respects, characteristic of the season. Their stalks are woody; but they are not shrubs, and their flowers are more delicate than brilliant. The foreign asters which are cultivated in our gardens, though exceeding the native species in the brilliancy of their hues, are inferior to the latter in elegance of growth, and in the delicate structure of their blossoms. The prevailing color of the autumnal flowers is yellow; yet there is not a single yellow aster among their whole extensive tribe. Near the latter part of September the fields are covered with asters of every shade, from the deep blue of the cyaneus and the purple of the New England aster, to the purest white. The walls and the edges of the woods are bordered with long rows of goldenrods, and multitudes of gaudy flowers have usurped the dominion of the roses, hiding the summer shrubbery beneath their tall and spreading herbage.

Some flowers are interesting because they are rare; some because they are common and familiar; some attract our attention because they are large, others because they are small. We are in the habit of admiring opposite qualities in very similar things. The pineweed, a little plant that is abundant in September, is interesting on account of its minuteness. We meet it in dry pastures and on rustic roadsides, on a thin and sandy soil, in company with the Trichostema, a very pretty annual with numerous blue flowers, each having two long stamens overarching the flower, so as to resemble *blue curls*, the common name of the species. The pineweed never fails to attract attention by its multitudes of little starlike flowers covering the plant like golden spangles. When the flower has perished it is succeeded by a diamond-shaped red capsule, so that the plant is as pretty with its red fruit as with its yellow flowers.

Almost simultaneously with the tinting of the forest-trees comes forth the last beautiful visitant of our fields, the blue-fringed gentian. This little flower marks the decline of autumnal vegetation. It begins to unfold itself during the latter part of September, and may often be found in the meadows after the November frosts have seared the verdure of the fields, and changed the variegated hues of the forest into one monotonous tinge of brown and purple.

When the woods are completely divested of their foliage, and the landscape wants nothing but snow to yield it the aspect of winter, the hamamelis, or witch-hazel, still retains its yellow blossoms, in defiance of the later frosts. Nothing is lively around it but the evergreens, and no plant puts forth its blossoms after this, unless some flower of spring should peep out unseasonably from under the protection of a sunny knoll. The evergreens are now in all their beauty, and we search the fields in vain for aught but the presages of winter.

# SEPTEMBER.

WE have hardly become familiar with summer ere autumn arrives with its cool nights, its foggy mornings, and its clear brilliant days. Yet the close of summer is but the commencement of a variety of pleasant rural occupations, of reaping and fruit-gathering, and the still more exciting sports of the field. After this time we are comparatively exempt from the extremes of temperature, and we are free to ramble at any distance, without exposure to sudden showers, that so often spring up in summer without warning us of their approach. Though the spicy odors of June are no longer wafted upon the gales, there is a clearness and freshness in the atmosphere more agreeable than fragrance, giving buoyancy to the mind and elasticity to the frame.

The various employments of the farmer are changed into agreeable recreations; and the anxious toils of planting and haymaking have given place to the less wearisome and more exhilarating labors of the harvest. Beside the pleasures of the sportsman, there are successions of fruit-gatherings and rural excursions of various kinds, from the beginning of this month to the end of the next, that impart to the young many cheerful themes for remembrance during the rest of their days. The provident simpler may be seen upon the hills busily employed in gathering medicinal plants for her own humble dispensary. Close by her side are neatly bound sheaves of thoroughwort, hardhack, bear-berry, pennyroyal, and life-everlasting, which she benevolently provides for the supply of her neighborhood. And while thus employed, she feels the reward of the just in the pleasing contemplation of the good she may perform, when winter comes with its fevers and colds.

There is no season when the landscape presents so beautiful an appearance just before sunset, as during this month. The grass has a singular velvety greenness, being without any mixture of downy tassels and panicles of seeds. For the present covering of the fields is chiefly the second growth of vegetation, after the first has been mowed by the farmer or cropped by the grazing herds. The herbage displays little but the leaves, which have been thickened in their growth and made green by the early rains of autumn. When the atmosphere has its usual autumnal clearness and the sun is just declining, while his rays gleam horizontally over the fields, the plain exhibits the most brilliant verdure, unlike that of the earlier months. When this wide landscape of uniform greenness is viewed in opposition to the blue firmament, it seems as if the earth and the sky were vying with each other in the untarnished loveliness of their appropriate colors.

There is usually a serenity of the weather for the greater part of September, unknown to the other autumn months. Yet this is no time for inaction; for the temperate climate, too pleasant for confinement, and too cool for indolent repose, invites even the weary to ramble. Of all the months, the climate of September is the most equable and salubrious, and nearly the same temperature is wafted from every quarter of the heavens. The sea-breezes spring up from the ocean almost with the mildness of the southwest, and the rude north-wind has been softened into a delightful blandness by his tender dalliance with summer.

One of the charms of the present month is the profusion

of bright-colored fruits that meet the eye on every side in the deserted haunts of the flowers. The scarlet berries of the nightshade, varied with their blossoms, hang like clusters of rubies from the crevices in the stone-walls through which the vines have made their clambering tour. On each side of the fences the elder-trees in interrupted rows are bending down with the weight of their dark purple fruit, and the catbird may be seen busily gathering them for his noonday repast. Above all, the barberry-bushes scattered over the hills, some in irregular clumps, others following the lines of the stone-walls, down narrow lanes and over sandy hills, with their long slender branches fringed with delicate racemes of variegated fruit, changing from a greenish white to a bright scarlet, form hedge-rows as beautiful as art, without its formality.

September is the counterpart of June, and displays the transformation of the flowers of early summer into the ripe and ruddy harvest. The wild-cherry trees are heavily laden with their dark purple clusters, and flocks of robins and waxwings are busy all the day in their merry plunder among the branches. But in the fruits there is less to be loved than in the flowers, to which imagination is prone to assign some moral attributes. The various fruits of the harvest we prize as good and bounteous gifts. But flowers win our affections, like beings endowed with life and thought; and when we notice their absence or their departure we feel a painful sense of melancholy, as when we bid adieu to living friends. With flowers we associate the sweetness, the loveliness, and the dear and bright remembrances of spring. Like human beings, they have contributed to our moral enjoyments. But there are no such ideas associated with the fruits, and while the orchards are resplendent with their harvest, they can never affect the mind like the sight of flowers.

The birds are almost silent; now and then we hear one piping a few broken strains, but he does not seem to be pleased with his own song, and no one answers him from his feathered comrades. Their season of departure is near, and numerous cares distract the tuneful band. The swallows· are no longer seen with twittering flight skimming along the surface of the waters, or sailing aloft in the air to warn the swain of coming showers. The little busy wren, one of our latest warblers, is also silent, and all are slowly leaving us one after another. It is a pleasant occupation to watch their various movements, their altered manners, and their unwonted shyness. They sing no more, but twitter, chirrup, and complain, always in motion, flying from tree to tree, and busy like those preparing for a long journey.

But as the birds have become silent, the insect myriads, having attained the maturity of their lives, are in glad chorus with all their little harps. The fields are covered with crickets and grasshoppers, and the whole air resounds with their hissing melodies. This is the honeymoon of their transient lifetime, and they are merrily singing their conjugal ditties, while the autumnal frosts are rapidly approaching to put an end to their pleasures and their lives. While chirping night and day among the green herbage, they are but chanting the deathnotes of their own brief existence. The little merry multitude, to whose myriad voices we are now listening with delight, contains not one individual of those who were chirping in their places a year ago. All that generation has passed away, and ere another spring arrives, the present multitudinous choir will have perished likewise, to yield their places to new millions, which the next summer will usher into life. But they take no thought of the morrow, and like true Epicureans, while the frosts are gathering around them, they sing and make merry until the cold drives them into their retreats. One tribe after another discontinues its song, until the hard frosts arrive, and leave the woods lonely and silent, but for the screaming of jays, the cawing of ravens, and the moaning of the winds as they pass over the graves of the departed things of summer.

UNIV.
CALIF.

# BIRDS OF THE NIGHT.

NUMEROUS swarms of insects and many small quadrupeds that require darkness for their security come abroad only during the night or twilight. These creatures would multiply almost without check, were it not that certain birds, having the power of seeing in the dark, and being partially blinded by daylight, are forced to seek their food in the night. Many species of insects, not strictly nocturnal, — those in particular that pass their life chiefly in the air, — are most active after dewfall. Hence the very late hour at which certain species of Swallows retire to rest, the period of sunset and early twilight affording them a fuller repast than any other part of the day. No sooner has the Swallow gone to rest than the Night-Jar and Whippoorwill come forth to prey on the larger kinds of aerial insects. The bat, an animal of antediluvian type, comes out a little earlier, and assists in lessening these multitudinous swarms. The small Owls, though they pursue the larger beetles and moths, direct their efforts chiefly at the small quadrupeds that steal out in the twilight to nibble the tender herbs and grasses. Thus, the night, except the hours of total darkness, is with many species of animals, though they pursue their objects with great stillness and silence, a period of general activity.

The birds of the night may be classed under two heads, including, beside the true nocturnal birds, that go abroad in the night to seek their subsistence, those diurnal birds that continue their songs. There are other species that are quiet both at noonday and midnight. Such is the Chimney-Swallow. This bird employs the middle of the day in sleep after excessive activity from the earliest dawn. It is seen afterwards circling about at the decline of day, and is sometimes abroad in fine weather the greater part of the night, when the young require almost unremitted exertions on the part of the old birds to procure their subsistence.

The true nocturnal birds, of which the Owl and the Whippoorwill are prominent examples, are distinguished by a peculiar sensibility of the eye that enables them to see clearly by twilight and in cloudy weather, while they are dazzled by the broad light of day. Their organs of hearing are proportionally delicate and acute. Their wing-feathers have a peculiar downy softness, so that they move through the air without the usual fluttering sounds that attend the flight of other birds. Hence they are able to steal unawares upon their prey, and to make their predal excursions without disturbing the general silence of the hour. This noiseless flight is remarkable in the Owl, as may be observed if a tame one is confined in a room, when we can perceive his motions only by our sight. It is remarkable that this peculiar structure of the wing-feathers does not exist in the Woodcock, which is a nocturnal feeder. Nature makes no useless provisions for her creatures. Hence this bird, that obtains its food by digging into the ground and takes no part of it while on the wing, has no need of such a contrivance. Neither stillness nor stealth would assist him in digging for his helpless prey.

### THE OWL.

Among the nocturnal birds the most celebrated is the Owl, of which there are many species, varying from the size of an Eagle down to the Acadian, which is no larger than a Robin. The resemblance of the Owl to the feline race has been a frequent subject of remark. Like the cat, he sees most clearly by twilight or the light of the moon, seeks his prey in the night, and spends the greater part of the day in sleep. This likeness is made stronger by his earlike tufts of feathers, that correspond with the ears of a quadruped; by his large head; his round, full, and glaring eyes, set widely apart; by the extreme contractility of the pupil; and by his peculiar habit of surprising his victims by watchfulness and stealth. His eyes are partially encircled by a disk of feathers, giving a remarkably significant expression to his face. His hooked bill, turned downwards so as to resemble the nose in the human face, the general flatness of his features, and his upright position produce a grave and intelligent look. It was this expression that caused him to be selected by the ancients as the emblem of wisdom and to be consecrated to Minerva.

The Owl is remarkable for the acuteness of his hearing, having a large ear-drum and being provided with an apparatus by which he can exalt this faculty when he wishes to listen with great attention. Hence, while he is noiseless in his own motions, he is able to perceive the least sound from the motion of any other object, and overtakes his prey by coming upon it in silence and darkness. The stillness of his flight adds mystery to his character, and assists in making him an object of superstitious dread. Aware of his defenceless condition in the bright daylight, when his purblindness would prevent him from evading the attacks of his enemies, he seeks some secure retreat where he may pass the day unexposed to observation.

It is this necessity which has caused him to make his abode in desolate and ruined buildings, in old towers and belfries, and in the crevices of dilapidated walls. In these places he hides from the sight of other birds, who regard him as a common enemy, and who show him no mercy when they have discovered him. Here also he rears his offspring, and we associate his image with these solitary haunts, as that of the Loon with our secluded lakes. In thinly settled and wooded countries, he selects the hollows of old trees and the clefts of rocks for his retreats. All the smaller Owls, however, seem to multiply with the increase of human population, subsisting upon the minute animals that accumulate in outhouses, orchards, and fallows.

When the Owl is discovered in his hiding-place, the alarm is given, and there is a general excitement among the small birds. They assemble in great numbers, and with loud chattering assail and annoy him in various ways, and soon drive him out of his retreat. The Jay, commonly his first assailant, like a thief employed as a thief-taker, attacks him with great zeal and animation. The Chickadee, the Nuthatch, and the Red-thrush peck at his head and eyes, while other birds less bold fly round him, and by their vociferation encourage his assailants and increase the terror of their victim.

It is while sitting on the branch of a tree or on a fence after his misfortune and escape that he is most frequently seen in the daytime. Here he has formed a subject for painters, who have generally introduced him into their pictures as he appears in one of these open situations. He is sometimes represented ensconced in his own select retreat, apparently peeping out of his hiding-place and only half concealed; and the discovery of him in such lonely places has caused the supernatural horrors attached to his image. His voice is supposed to bode misfortune, and his spectral visits are regarded as the forerunners of death. His occupancy of deserted houses and ruins has invested him with a romantic character, while the poets, by introducing him to deepen the force of their pathetic or gloomy descriptions, have enlivened our associations connected with his image; and he deserves therefore in a special degree to be classed among those animals which we call picturesque.

Though the Owl was selected by the ancients as the emblem of wisdom, the moderns have practically renounced this idea, which had its foundation in the gravity and not in the real character of the bird, which possesses only the sly and sinister traits that mark the feline race. A very different train of associations and a new series of picturesque images are now suggested by the figure of the Owl, who has been more correctly portrayed by modern poetry than by ancient mythology. He is now universally regarded as the emblem of ruins and of desolation, — true to his character and habits, which are intimately allied with this description of scenery.

I will not enter into a speculation concerning the nature and origin of those agreeable emotions which are so generally produced by the sight of objects that suggest ideas of ruins. It is happy for us that by the alchemy of poetry we are able to turn some of our misfortunes into sources of melancholy pleasure, after the poignancy of grief has been assuaged by time. Nature has also benevolently provided that many an object that is capable of communicating no direct pleasure to our senses shall affect us agreeably through the medium of sentiment. Thus, the image of the Owl awakens the sentiment of ruin; and to this feeling of the human soul we may trace the pleasure we derive from the sight of this bird in his appropriate scenery. Two Doves upon the mossy branch of a tree, in a wild and beautiful sylvan retreat, are the pleasing emblems of love and constancy; but they are not more suggestive of poetic fancies than an Owl sitting upon an old gate-post near a deserted house.

I have alluded in another page to the faint sounds we

hear when the birds of night, on a still summer evening, are flying over short distances in a neighboring wood. There is a feeling of mystery awakened by these sounds that exalts the pleasure we derive from the delightful influence of the hour and the season. But the emotions thus produced are of a cheerful kind, slightly imbued with sadness, and not equal in intensity to the effects of the hardly perceptible sound occasioned by the flight of the Owl as he glides by in the dusk of evening or in the dim light of the moon. Similar in effect is the dismal voice of this bird, which is harmonized with darkness, and, though in some cases not unmusical, is tuned as it were to the terrors of that hour when he makes secret warfare upon the sleeping inhabitants of the wood.

### THE ACADIAN OWL, OR SAW-WHETTER.

One of the most interesting of this family of birds is the little Acadian Owl, whose note formerly excited much curiosity. In the " Canadian Naturalist" an account is given of a rural excursion in April, when the attention of the party was called, just after sunset, to a peculiar sound heard in a cedar-swamp. It was compared to the measured tinkling of a cowbell, or to regular strokes upon a piece of iron quickly repeated. One of the party, who could not describe the bird, remembered that " during the months of April and May, and in the former part of June, we frequently hear after nightfall the sound just described. From its regularity it is thought to resemble the whetting of a saw, and hence the bird from which it proceeds is called the Saw-Whetter."

These singular sounds are the notes of the Acadian Owl. They are like the sound produced by the filing of a mill-saw, and are said to be the amatory note of the male, being heard only during the season of incubation.

Mr. S. P. Fowler informed me by letter that "the Acadian Owl has another note which we frequently hear in the autumn after the breeding season is over. The parent birds, then accompanied by their young, while hunting their prey in the moonlight, utter a peculiar note resembling a suppressed moan or low whistle. The little Acadian, to avoid the annoyance of the birds he would meet by day, and the blinding light of the sun, retires in the morning, his feathers wet with dew and rumpled by the hard struggles he has encountered in seizing his prey, to the gloom of the forest or the thick swamp. There, perched on a bough near the trunk of the tree, he sleeps through a summer's day, the perfect picture of a *used-up* little fellow, suffering the evil effects of a night's carouse."

### THE SCREECH-OWL.

The Mottled Owl, or Screech-owl, is somewhat larger than the Acadian, or Whetsaw, but not so familiar as the Barn Owl of Europe, which he resembles. He builds in the hollows of old trees and in deserted buildings, whither he resorts in the daytime for repose and security. His voice is heard most frequently in the latter part of summer, when the young owlets are abroad. They use their cries for mutual salutation and recognition. The wailing note of this Owl is singularly wild and not unmusical. It is not properly a screech or a scream, like that of the hawk or the peacock, but rather a sort of moaning melody, half music and half bewailment. This plaintive strain is far from disagreeable, though it has a cadence expressive of dreariness and desolation. It might be performed on a fife, beginning with D octave and running down by quarter-tones to a third below, frequently repeating the notes with occasional pauses for about one minute. The bird does not slur his notes, but utters them with a sort of tremulous staccato. The separate notes may be distinctly perceived, though the intervals are hardly appreciable.

The generality of this family of birds cannot be regarded as useful. They are only mischievous birds of prey, and no more entitled to mercy or protection than the Falcons, to which they are allied. All the little Owls, however, though guilty of destroying small birds, are serviceable in ridding our fields and premises of mischievous animals. They destroy multitudes of large nocturnal insects, flying above the summits of trees in pursuit of them, while at other times their flight is low, when watching for mice and moles, that run upon the ground. It is on account of its low flight that the Owl is seldom seen upon the wing. Bats, which are employed by Nature for similar services, fall victims in large numbers to the Owls, which are the principal means of checking their multiplication.

An interesting family of nocturnal birds are the Moth-hunters, of which in New England there are only two species, the Whippoorwill and the Nighthawk. These birds resemble the Owls in some of their habits; but in their structure, their mode of obtaining subsistence, and in their general characters they resemble Swallows. They are shy and solitary, take their food while on the wing, abide chiefly in the deep woods, and come abroad only at twilight or in cloudy weather. They remain, like the Dove, permanently paired, lay their eggs on the bare ground, and, when perched, sit upon the branch lengthwise, unlike other birds. They are remarkable for their singular voices, and only one species — the Whippoorwill — may be considered musical. They are inhabitants of all parts of the world, but are particularly numerous in the warmer regions of North and South America, where the curiosity of the traveller is constantly excited by their voices resembling human speech.

## THE WHIPPOORWILL.

The Whippoorwill is well known to the inhabitants of New England by his nocturnal song. This is heard chiefly in wooded and retired situations, and is associated with the solitude of the forest as well as the silence of the night. The Whippoorwill is therefore emblematic of the rudeness of primitive nature, and his voice reminds us of seclusion and retirement. Sometimes he wanders away from the wood into the precincts of the town, and sings near our dwelling-houses. Such an incident was formerly the occasion of superstitious alarm, and was regarded as an omen of evil to the inmates of the dwelling. The cause of these irregular visits is probably the accidental abundance of a particular kind of insects which the bird has followed from the woods.

The Whippoorwill in this part of the country is first heard in May, and continues vocal until the middle of July. He begins to sing at dusk; and we usually hear his note soon after the Veery, the Philomel of our summer evenings, has become silent. His song consists of three notes, in a sort of polka-time, with a slight rest after the first note in each bar, as given below: —

Whip poor will  Whip poor will  Whip poor will  Whip.

I should remark that the bird begins his song with the second syllable of his name, if we may suppose him to utter the word, or I might say with the second note in the bar. Some birds occasionally, though seldom, fall short of these musical intervals, as they are written on the scale, and an occasional cluck is heard when we are near the singer. The notes of the Quail so clearly resemble those of the Whippoorwill that I give them below, that they may be compared.

So great is the similarity of the notes of these two birds, that those of the Quail need only be repeated in succession without pause to be mistaken, if heard in the night, for those of the Whippoorwill. They are uttered with a similar intonation; but the voice of the nocturnal bird is more harsh, and his song consists of three notes instead of two, and is pitched a few tones higher.

The song of the Whippoorwill, though wanting in mellowness of tone, as may be perceived when we are near him, is very agreeable except to a few, notwithstanding the superstitions associated with it. Some persons are not disposed to class the Whippoorwill among singing-birds, regarding him as more vociferous than musical. But it would be difficult to determine in what respect his notes differ from the songs of other birds, except that they approach more nearly to the precision of artificial music. Yet it will be admitted that a considerable distance is required to "lend enchantment" to the sound of his voice. In some retired and solitary districts, the Whippoorwills are so numerous as to be annoying by their vociferations. But in those places where only a few individuals are heard during the season, their music is a source of great pleasure, and constitutes one of the principal charms of the neighborhood.

I was witness of this some years ago, in one of my botanical rambles in Essex County, which is for the most part too open and cleared to suit the habits of these solitary birds. On one of these excursions, after walking several hours over a rather wild region I arrived at a very romantic spot, consisting of an open level, completely surrounded by woods. Nature uses her ordinary materials to form her most delightful landscapes, and causes them to rise up as it were by magic when we least expect them. Here I suddenly found myself encompassed by a charming amphitheatre of hills and woods, and in a valley so beautiful that I could not have imagined anything equal to it. A neat cottage stood with only one other in this spot. It was entirely wanting in any architectural decoration, which I am confident would have dissolved the spell that made the whole scene so attractive. It was occupied by a shoemaker, whom I recognized as an old acquaintance and a worthy man, who resided here with his wife and children, whose mode of living was one of the few vestiges of ancient simplicity. I asked them if they were contented while living so far from the town. The wife of the cottager replied that they suffered in the winter from their solitude; but in the warm season they preferred it to the town, "for in this place we hear all the singing-birds early and late, and the Whippoorwill sings every night during May and June." It was the usual habit of this bird, they told me, to sing both in the morning and evening twilight; but if the moon should rise late in the evening after it had become silent, it would begin to sing anew as if to welcome her rising. May the birds continue to sing to this happy family, and may the voice of the Whippoorwill never bode them any misfortune!

### THE NIGHT-JAR.

The Night-Jar, or Nighthawk, is similar in many points to the Whippoorwill. The two, indeed, were formerly considered identical; but more careful investigation has proved them to be distinct species. I believe that some extraordinary pedant has also demonstrated that they belong to two distinct genera. Let us take heed that science do not degenerate, like metaphysics, into a mere vocabulary of distinctions which only the mind of a Hudibras can appreciate. The two birds, however, are not identical. The Nighthawk is a smaller bird, has no song, and exhibits many of the ways of the Swallow. He is marked by a white spot on his wings, which is very apparent during his flight. He seems to take his prey in a higher region of the atmosphere, being frequently seen, at twilight and in cloudy weather, soaring above the house-tops in quest of insects. The Whippoorwill finds his subsistence chiefly near the ground, flitting about the farmyard, the fences, and wood-piles, and taking an insect from a branch of a tree, while poising himself on the wing like a Humming-Bird. He is never seen circling aloft like the Nighthawk.

The movements of the Nighthawk during his flight are performed generally in circles, and are very picturesque. The birds are usually seen in pairs at such times, but occasionally there are numbers assembled together; and one might suppose they were engaged in a sort of aerial dance, and that they were emulating each other in their attempts at soaring to a great height. It is evident that these evolutions proceed in part from the pleasure of motion, but they are also a few of their ways during courtship. While they are soaring and circling in the air, they occasionally utter a shrill note which has been likened to the word *Piramidig*, forming a name by which the bird is sometimes called. Now and then they are seen to dart with a rapid motion to take a passing insect.

While performing these circumvolutions, the male occasionally dives perpendicularly downwards, through a considerable space, uttering, as he makes a sudden turn upwards at the bottom of his descent, a singular note resembling the twang of a viol-string. This sound has been supposed to be made by the action of the air as the bird dives swiftly through it with open mouth. This is proved to be an error by the fact that the European species makes a similar sound while sitting on its perch. Others think that this diving motion of the bird is designed to intimidate those who seem to be approaching his nest; but the bird performs the same manœuvre when he has no nest to defend. This habit is peculiar to the male, and it is probably one of those fantastic motions which are noticed among the male Doves as artifices to attract the attention of the female.

This twanging note, made during the precipitate descent of the Nighthawk through the air, is one of the picturesque sounds of Nature, and is heard most frequently in the morning twilight, when the birds are collecting their early repast of insects. If we should go abroad before daylight or at the earliest dawn, we might see them circling about, and hear their cry frequently repeated. Suddenly this twanging sound excites our attention, and if we were not acquainted with it or with the habits of the bird, we should feel a sensation of mystery, for there seems to be nothing like it in nature. The sound produced by the European species is a sort of drumming or whizzing note, like the hum of a spinning-wheel. The male begins this performance about dark, and continues it at intervals a great part of the night. It is effected while the breast is inflated with air, like that of a cooing Dove. The Nighthawk inflates its breast in a similar manner, and utters a similar sound when any one approaches the nest.

The habit of the Whippoorwill and Nighthawk of sitting lengthwise and not crosswise on their perch has excited some curiosity; for it is well known that these birds are capable of grasping a perch and sitting upon it. On the contrary, they roost upon a large and nearly horizontal branch in a longitudinal direction. The design of nature in this instinct is to afford the bird that concealment which is needful for its protection in the daytime. When thus placed, he is entirely hidden from sight below. The Owl is protected by another mode of concealment. He sits very erect, near the bole of the tree, and draws his tail-feathers right against the branch, so that he can hardly be seen from below. The Nighthawk, while reposing lengthwise upon his perch, would, if his foe were looking down upon him, hardly be distinguished when his mottled-brown plumage made no contrast in color with the bark of the tree.

### THE MOCKING-BIRD.

I will now turn my attention to those diurnal birds that sing in the night as well as in the day, and are classed under the general appellation of Nightingales. These birds do not confine their singing to the night, like the Whippoorwill, and are most vocal by twilight and the light of the moon. Europe has several of these minstrels of the night, beside the true Philomel of poetry and romance. In the United States the Mocking-Bird enjoys the greatest reputation; but there are other birds of more solitary habits and less known, among which are the Rose-breasted Grosbeak and the Water-Thrush, that sing in the night.

The Mocking-Bird is well known in the Middle and Southern States, but seldom passes a season in New England, except in the southern extremity, which seems to be the limit of its northern residence. Probably like the Grosbeak, which is constantly extending its range in an eastern direction, the Mocking-Bird may be gradually making progress northwardly; so that fifty years hence each of these birds may be common in the New England States. The Mocking-Bird is familiar in his habits, frequenting gardens and orchards, and perching on the roofs of houses when singing, like the common Robin. Indeed, this bird owes much of his popularity to his familiar and amiable habits. Like the Robin, too, a bird that sings at all hours except those of complete darkness, he is a persevering songster, and seems to be inspired by living in the vicinity of man. In his manners, however, he bears more resemblance to the Red-Thrush, being distinguished by his vivacity and his courage in repelling the attacks of his enemies.

The Mocking-Bird is celebrated throughout the world for his musical powers; but it is difficult to ascertain precisely the character and quality of his original notes. Some naturalists affirm that he has no notes of his own, but confines himself to imitations. That this is an error, all persons who have listened to his native wild notes can testify. I should say, from my own observations, not only that he has a distinct song, peculiarly his own, but that his best imitations will bear no comparison with his native notes. His common habit during the day is to utter frequently a single strain, hardly distinguishable from that of the Red-Bird, and similar to that of the Baltimore Oriole. This seems to be his amusement while busy with the affairs of his own household and providing for their wants. It is only when confined in a cage that he is constant in his mimicry. In his native woods, and especially at an early hour in the morning, when he is not provoked to imitation by the notes of other birds and animals, he sometimes pours out his own wild notes with uninterrupted fervor. Yet I have often listened vainly for hours to hear him utter anything more than a few idle repetitions of monotonous sounds, interspersed with some ludicrous variations. Why he should discard his own delightful song to tease the listener with all imaginable discords is not easily explained.

Though his powers of mimicry are the cause of his fame, his real merit is not based upon these. He would be infinitely more valuable as a songster, if he were incapable of imitating a single sound. I would add that as an imitator of the songs of other birds he is very imperfect, and has been greatly overrated by our ornithologists, who seem to vie with each other in their exaggerations of his powers. He cannot utter correctly the notes of the rapid singers. He is successful only in his imitations of those birds whose notes are simple and moderately delivered. Hence he gives good imitations of the Robin. He is, indeed, more remarkable for his indefatigable propensity than for his powers, in which he is exceeded by some Parrots. Single sounds, from whatever source they may come, — from birds, quadrupeds, reptiles, or machines, — he delivers very accurately. But I have heard numbers of Mocking-Birds in confinement attempt to imitate the Canary without success. There is a common saying that the Mocking-Bird will die of chagrin if placed in a cage by the side of a caged Bobolink, mortified because he cannot give utterance to his rapid notes. If this would cause his death, he would also die when confined near a Canary or with any of the rapid-singing Finches. It is also an error to say of his imitations, as writers assert, that they are improvements upon the originals. When he utters the notes of the Red-Bird, the Oriole, or the common Robin, his imitations are perfect, but are no clearer or sweeter; and when he gives us the screaming of a Jay, the mewing of a cat, or the creaking of a cart-wheel, he does not change them into music.

As an original songster, estimated by the notes which on rare occasions he pours out in a serious mood from his own favorite spot and during his favorite hour, which is the earliest dawn, the Mocking-Bird is probably unequalled by any American songster. His notes are loud, varied, melodious, and of great compass. They may be likened to those of the Red-Thrush, more forcibly delivered, and having more flute-notes and fewer guttural notes and sudden transitions. He also sings often on the wing, and with fervor, while the other Thrushes sing only from their perch. But his song has less variety than that of the Red-Thrush, and falls short of it in some other respects. The Red-Thrush, however, has too little persistence in his singing.

By other writers the Mocking-Bird is put forward as superior to the Nightingale. This assumption might be worthy of consideration, if the American bird were not addicted to mimicry. This execrable habit renders him unfit to be compared with the Nightingale, whose song also resembles that of a Finch more than that of a Thrush. His mocking habits almost annihilate his value as a songster; as the effect of a concert would be spoiled if the players were constantly introducing, in the midst of their serious music, snatches of vulgar and ridiculous tunes and uncouth sounds.

## TO THE MOCKING-BIRD.

CAROLLING bird, that merrily night and day
Tellest thy raptures from the rustling spray,
And wakest the morning with thy varied lay,
  Singing thy matins ; —
When we have come to hear thy sweet oblation
Of love and joyance from thy sylvan station,
Why in the place of musical cantation
  Balk us with pratings ?

# BIRDS OF THE NIGHT.

We stroll by moonlight in the dusky forest
Where the tall cypress shields thee, fervent chorist!
And sit in haunts of echoes when thou pourest
    Thy woodland solo.
Hark! from the next green tree thy song commences;
Music and discord join to mock the senses,
Repeated from the tree-tops and the fences,
    From hill and hollow!

A hundred voices mingle with thy clamor;
Bird, beast, and reptile take part in thy drama;
Outspeak they all in turn without a stammer, —
    Brisk Polyglot!
Voices of kill-deer, plover, duck, and dotterel;
Notes, bubbling, hissing, mellow, sharp, and guttural,
Of catbird, cat, or cart-wheel, thou canst utter all,
    And all untaught.

The raven's croak, the chirrup of the sparrow,
The jay's harsh note, the creaking of a barrow,
The hoot of owls, all join the soul to harrow
    And grate the ear.
We listen to thy quaint soliloquizing,
As if all creatures thou wert catechizing,
Tuning their voices, and their notes revising
    From far and near.

Sweet bird, that surely lovest the "noise of folly,"
Most musical, but never melancholy;
Disturber of the hour that should be holy,
    With sounds prodigious; —
Fie on thee! O thou feathered Paganini!
To use thy little pipes to squawk and whinny,
And emulate the hinge and spinning-jenny,
    Making night hideous.

Provoking melodist! why canst thou breathe us
No thrilling harmony, no charming pathos,
No cheerful song of love, without a bathos?
    The Furies take thee!
Blast thy obstreperous mirth, thy foolish chatter, —
Gag thee, exhaust thy breath, and stop thy clatter,
And change thee to a beast, thou senseless prater!
    Naught else can check thee!

## BIRDS OF THE NIGHT.

A lengthened pause ensues ; but hark again !
From the near woodland, stealing o'er the plain,
Comes forth a sweeter and a holier strain !
      Listening delighted,
The gales breathe softly, as they bear along
The warbled treasure, the tumultuous throng
Of notes that swell accordant in the song,
      As love is plighted.

The echoes, joyful, from their vocal cell,
Leap with the wingéd sounds o'er hill and dell,
With kindling fervor as the chimes they tell
      To wakeful even :
They melt upon the ear ; they float away,
They rise, they sink, they hasten, they delay,
And hold the listener with bewitching sway,
      Like sounds from heaven.

# RUINS.

To all whose minds have received an ordinary amount of cultivation there are few objects more interesting than the remains of antiquity, — whether, like those of Greece and Rome, they call up the history of the noblest works of art and deeds of renown, and like those of Egypt, carry back the mind to the age of primeval superstition, or, like the ruins of the earth itself, they repeat the story of the antediluvian periods, before the present races of animals appeared. In our own country where these relics of ancient times, excepting those of a geological description, are almost unknown, the people in general can hardly sympathize with that love of ruins which is almost a passion with some of the inhabitants of the Old World. We have no ruined castles to remind us of ancient baronial splendor, and of the perils and heroism of the feudal ages; no remains of gorgeous temples or triumphal arches, to record the deeds of a past generation. The ancient history of this continent lives chiefly in tradition; and the traveller who happens to discover one of the few relics of ancient American architecture seeks in vain for any record that will explain its character or design.

Yet the absence of the ruins of antiquity may have a tendency to render our people more alive to impressions from those of a humble description and of recent origin which abound in all places. When strolling over the scenes of our own land, who has not often stopped to ponder over the ruins of some old dwelling-house, and to bring before the mind the possible history of its inmates? There we perceive the completion of a domestic romance. A series of adventures has been there commenced, continued, and brought to an end. Imagination is free to indulge itself in making up the history of the human beings who have lived and died there, and of the romantic adventures which have been connected with it. We do not always endeavor to read this history; but there is a shadowy conception of something associated with the old crumbling walls that would be striking and romantic. To this pleasing occupation of the fancy may undoubtedly be ascribed a portion of the interest always excited by a view of a ruined or deserted house. A still deeper effect is produced by the sight of a mouldering temple or a ruined castle, which are allied with deeds and events of greater magnitude.

I am disposed to attribute the pleasure arising from the contemplation of ruins to an exalted affection of the human soul, to a veneration of the past, and a longing to recover the story of bygone ages. A ruin is delightful as the scene of some old tradition, a specimen of ancient art and magnificence, and as evidence of the truth of history. Nothing, indeed, serves to place so vividly before the mind the picture of any historic event as the ivied and dilapidated walls of the building in which it occurred. There is likewise an emotion of cheerful melancholy which is awakened by viewing a pile of ruins, an old house, or an old church, venerable with the mosses of time and decay. There are other objects, scenes, and situations that produce similar effects upon the mind, such as a sight of the ocean when agitated by a tempest, from a place of security. A beacon and a lighthouse belong to the same class of objects; and above all, a monument by the sea-shore, erected to commemorate some remarkable shipwreck, awakens a train of melancholy reflections nearly allied to the sentiment of ruins. But it is not every scene of ruins that is capable of yielding pleasure to the beholder. There is nothing agreeable in a view of the embers of a wide conflagration, except a gratification of the curiosity. Such a spectacle brings to the mind only the idea of dissolution and misfortune, which is painful, and there is nothing connected with it to awaken any counteracting sentiment. On the other hand, every mind is agreeably affected by the sight of an old house, no longer the habitation of man, serving only as the day retreat of the owl and the fancied residence of beings of the invisible world. There is a propensity among men to associate every ruined edifice, however great or humble, with some romance or superstition; and our own people, who have no magnificent ruins, indulge the sentiment which is awakened by them in their legends of haunted houses, and by identifying these superstitions with every deserted habitation.

It is worthy of remark that although a cottage is more poetical than a palace, when each is in a perfect condition, a ruined palace is more poetical than a' ruined cottage. A certain amount of grandeur must be associated with a ruin to render it very effective. After a family have deserted their habitation of luxury and splendor, when they themselves have gone down to the grave, and their old mansion is crumbling with the ravages of time, we lose all that invidious feeling which often prevents us from sympathizing with the wealthy when they are living. They are now on a level with the humblest cottagers, and we look upon their ruined abode with a feeling of regret for all the elegance and greatness that have passed away. Indeed, the more noble and magnificent the edifice in its original state, the deeper is the emotion with which we contemplate its ruins. This circumstance yields a singular charm to the remains of the ancient Grecian temples, and to those Gothic castles that add a romantic character to certain European landscapes.

Some of the interesting accompaniments of a ruined building are the plants which are found clustering around its old roof and walls. Nature always decorates what time has destroyed, and when the ornaments of art have crumbled, she rears in their place garlands from her own wilds, and the building, no longer beautiful, is adorned with the greenness of vegetation. Hence certain plants have become intimately allied with ruins, and derive from this alliance a peculiarly romantic interest. Such are the mosses and lichens, the evergreen ferns, the creeper, and the most of the saxatile plants in America; in Europe, the yellow wall-flower, the chenopody, and the ivy.

In every ruin, therefore, we see the commencement of a new and beautiful creation. When a tree has fallen and has begun to decay, an infinite host of curious and delicate plants, of the simplest vegetable forms, are nurtured upon the surface of its trunk. Mushrooms of every description spring out from the inner bark, and lichens and mosses, as various in their hues as they are delicate in their forms, decorate all the outside. Insects which, under the magnifying-glass, exhibit the various plumes and glittering ornaments of the most brilliant birds and butterflies, live under the protection of these minute plants, as the larger animals find shelter in a forest of trees. When the tree has entirely perished, and has become assimilated with the soil, other hosts of plants of a higher order take the place of the former, until new forests have reared their branches over the ruins of those of a preceding age. Rocks, continents, and worlds are subject to the same decay and the same ultimate renovation. Thus the whole system of the universe is but an infinite series of permutations and combinations, all the atoms, amidst apparent chaos, moving in the most mathematical order, and gradually resolving themselves into organized forms, infinite in their numbers and arrangements.

In this country we have no classic ruins. The relics of the ancient structures of the aborigines can hardly awaken a romantic sentiment. We cannot associate with them any affecting historic reminiscences. We behold in them only the evidences of savage customs, unformed art, and a miserable superstition, which afford nothing to admire. No scenes are so well fitted as the ruins of a great and civilized nation, to inspire the mind with that contemplative habit which is the foundation of the poetical character. They fill the soul with noble conceptions, and serve to divert the thoughts from a consideration of present interest, and turn them back upon the ages of chivalry and romance.

Nature has so constituted the mind as to enable it to convert all her scenes, under certain circumstances, into sources of pleasure. It is not the beautiful alone that afford these agreeable impressions; nor is it the cheerful scenes only among natural or artificial objects that inspire a pleasing sentiment. While contemplating a scene of ruins, the mind may have glimpses of truths which are not revealed to us in the lessons of philosophy, and which excite indefinite hopes amidst apparent desolation. It is our power of deriving pleasure from these inexplicable sources that gives a pile of ruins half its charms. This mingled sentiment of hope and melancholy combines with almost all our ideas of beauty. On this account a deserted house interests the mind more than a splendid villa in its perfect condition; and a plain, overspread with classic ruins, more than a prospect of green meadows and highly ornamented gardens. It would be idle to assert that the human soul would take satisfaction in contemplating an object that is suggestive of its own dissolution. This love of ruins ought rather to be considered as so much evidence coming from them in favor of the infinite duration of the universe. They are evidence of the great age of the earth, and proof of its destination to exist during countless ages of the future. I wonder that our theologians have never deduced from this love of ruins, which is so universal, an argument for the immortality of the soul. It is evident that we do not instinctively regard them as proofs of mortality; but while we see in them the subjection of material forms to those changes which belong to everything that is mortal, we look upon our own souls as lifted above any liability to these changes. Did we innately perceive in them proof that the mind that constructed these wonderful works of art perished with them, we should turn away from them with a deep despondency, and endeavor to hide them from our sight. By a similar course of reasoning we may account for the pleasure which is experienced when musing among the tombs.

The scenes in our own land which are most nearly allied to ruins are the ancient rocks that gird our shores and give variety to our landscapes. They are, in fact, the ruins of an ancient world, existing probably before the human race had made their abode here. In these rocks the frosts of thousands of winters, and the lightnings of as many summers have made numerous fissures, and split them asunder in many places. We find the same species of saxatile and parasitic plants clustering about them which are found among the ruins of art. The forest-trees have inserted their roots into their crevices, and oaks that have stood for centuries nod their heads over the brink of these precipices, and cast a gloomier shade into the valleys below. Nothing can be more affecting than some of these ruins of nature, that want only the historical associations connected with the ruins of temples and palaces, to render them equally interesting.

Man's natural love of mystery, and his proneness to indulge in that emotion of grandeur and infinity that flows from the sight of anything involved in the dimness of remote ages of the past, are causes of the intense interest felt in the study of geology. With a deep feeling of awe we trace the footprints of those unknown animals which were the denizens of a former world. The mind " is roused to profound contemplation at the sight of piles of rocks as high as the clouds, recumbent on a bed of fern, and at finding the remains of animals that once sported on the summits of other Alps, now buried beneath the very base and foundation of ours."

# CALCULATIONS.

IT is remarkable that in this "enlightened age" (I give the quotation-marks, lest I might be suspected of originating the expression) there should be a necessity of entering upon a course of argument to prove the utility of birds to agriculture. It is also surprising that the greatest enemies of birds are among men whose occupation would be ruined if they were for a single year wholly deprived of their services. There are many who plead for the birds as beautiful and interesting objects, deserving protection for their own sake. But, valuable as they are for their songs, their gay plumage, and their amusing habits, all these qualities are of minor importance compared with the benefits they confer upon man, as checks to the overmultiplication of insects. The trees and the landscapes are made greener and the flowers more beautiful in the spring, the fruits of autumn finer and more abundant, and all nature is preserved in freshness and beauty by these hosts of winged musicians, who celebrate their garrulous revelries in field and wood.

I believe it admits of demonstration, that if birds were exterminated man could not live upon this earth. Almost every one of the smaller species is indispensable to our agricultural prosperity. The gunner who destroys ten small birds in the spring preserves as many millions of injurious insects to ravage our crops and render barren our orchards. Naturalists are unanimous in declaring the importance of their services; but cultivators, who of all persons ought to be most familiar with the facts that prove their usefulness, are indeed the most ignorant of them. They attribute to them a full moiety of the injury occasioned by insects; yet there is not an insect in existence which is not the natural food of certain birds, and which would multiply to infinity if not kept in check by them.

Men are willing and eager to keep dogs and cats, to feed and protect them, and endure their annoyances, because they understand that their services in a variety of ways, both in the house and out of doors, are sufficient to compensate for all their mischief and their trouble. They can appreciate their value, and are willing to overlook their offences. But the birds, who sing and make themselves agreeable in thousands of ways, men will destroy, because they are either too ignorant or too stupid to understand the benefits they derive from them. Probably the cats and dogs in this country cost in the aggregate a million of dollars in feeding them, to say nothing of their troublesomeness, to one hundred dollars which the whole feathered tribe costs us by the fruit and grain they damage and consume.

Calculations have been frequently made to ascertain the probable amount of insects consumed by any single bird. Many of these accounts are almost incredible, yet the most of them will admit of demonstration. Two different methods have been adopted for ascertaining these facts. The investigators watch the birds, to learn their food by their habits of feeding or foraging; or they destroy single birds at different times and seasons and examine the contents of their crop. Mr. Bradley, an English writer, mentions a person who was led by curiosity to watch a pair of birds that were raising a young brood, for one hour. They went and returned continually, bringing every time a caterpillar to the nest. He counted the journeys they made, and calculated that one brood did not consume less than five hundred caterpillars in the course of the day. The quantity destroyed in thirty days, at this rate, by one nest would amount to fifteen thousand. Suppose every square league of territory contained one hundred nests of this species, there would be destroyed by them alone in this space a million and a half of caterpillars in the course of one month.

I was sitting at a window one day in May, when my sister called my attention to a Golden Robin in a blackcherry tree employed in destroying the common hairy caterpillars that infest our orchards, and we counted the number he killed while he remained on the branch. During the space of one minute, by a watch, he destroyed seventeen caterpillars. I observed that he did not swallow the whole insect. After seizing it in his bill, he set his foot upon it, tore it asunder, and swallowed an atom taken from the inside. Had he eaten the whole caterpillar, three or four would probably have satisfied his appetite. But the general practice of birds that devour hairy caterpillars is to eat only a favorite morsel. Hence, they require a greater number to satisfy their wants.

This fact led me to consider how vast an amount of benefit this single species must contribute to vegetation. Suppose each bird to pass twelve out of the twenty-four hours in seeking his food, and that one hour of this time is employed in destroying caterpillars. At the rate of seventeen per minute, each bird would destroy a little more than one thousand caterpillars daily while they were to be found. Yet, if the crop of the bird were dissected, it would not be possible to discover from these titbits the character of the insect which he had devoured. So I draw the inference that while we may discover many important facts by dissection, all are not revealed to us by this mode of examination. Imagine, however, from the facts which I have recounted, the vast increase of cater-

T

pillars that would follow the extinction of this single species.

It is recorded in "Anderson's Recreations," that a curious observer, having discovered a nest of five young jays, remarked that each of these birds, while yet very young, consumed daily at least fifteen full-sized grubs of the May-beetle, and would require many more of a smaller size. The writer conjectures that of large and small each bird would require about twenty for its daily supply. At this rate the five birds would consume one hundred. Allowing that each of the parents would require fifty, the family would consume two hundred every day, and the whole amount in three months would be about twenty thousand. This seems to me from my own experience a very moderate calculation.

In obedience to an almost universal instinct, the granivorous birds, except those that lead their brood around with them like the hen, feed their young entirely upon the larva of insects. The finches and sparrows are therefore insectivorous, with but a few exceptions, the first two or three months of their existence. They do not consume grain or seeds until they are able to provide for themselves. The old birds supply their young with larva, when this kind of food is abundant, and when the tender state of their digestive organs requires the use of soft food. According to Mr. Augustus Fowler, who is good authority for any original observations, the American Goldfinch waits, before it builds a nest, until it is so late that the young, when they appear, may be fed with the milky grains and seeds of plants. It should be added that doves and pigeons soften the grain in their own crop before they give it to their young.

The quantity of insects consumed by the feathered race is infinite or beyond all calculation. The facts related of them show that birds require a larger quantity of food according to their size than quadrupeds. My own experience corroborates the accounts which I have selected from the testimony of other observers. I took from the nest two young bluebirds, which are only half the size of a jay, and fed them with my own hands for the space of two weeks. These little birds would each swallow twelve or more large muck-worms daily, or other grubs and worms in the same proportion. Still they always seemed eager for more, and were not overfed. I made a similar experiment with two young catbirds, which were attended with results still more surprising. Their voracity convinced me that the usual calculations bearing upon this subject are not exaggerated.

# WHY BIRDS SING IN THE NIGHT.

IN connection with this theme, we cannot escape a feeling of regret, almost like sorrow, when we reflect that the true nightingale and the skylark — the classical birds of European literature — are strangers to our fields and woods. In May and June there is no want of sylvan minstrels to wake the morn and to sing the vespers of a quiet evening. A flood of song awakens us at the earliest daylight; and the shy and solitary veery, after the vesper bird has concluded his evening hymn, pours his few pensive notes into the very bosom of twilight, and makes the hour sacred by his melody. But after twilight is sped and the moon rises to shed her meek radiance over the sleeping earth, the nightingale is not here to greet her rising, and to turn her melancholy beams into brightness and gladness. When the queen moon is on her throne, "clustered around by all her starry Fays," the whippoorwill alone brings her the tribute of his monotonous song, and soothes the dull ear of night with sounds which, however delightful, are not of heaven.

We have become so familiar with the lark and the nightingale by perusing the romance of rural life, that "neither breath of Morn when she ascends" without this the charm of her earliest harbinger, nor "silent Night" without her "solemn bird," seems holy as when we read of them in pastoral song. Poetry has hallowed to our minds the pleasing objects of the Old World. Those of the New must be cherished in song many more years before they can be equally sacred to the imagination.

## WHY BIRDS SING IN THE NIGHT.

The *cause* of the nocturnal singing of birds that do not go abroad during the night, and are strictly diurnal in all their other habits, has never been rationally explained. It is natural that the whippoorwill, which is a nocturnal bird, should sing during his hours of wakefulness and activity, and we may explain why ducks and geese, and other social birds, should utter their alarm-notes when they meet with any midnight disturbance. The crowing of a cock bears still more analogy to the song of birds; for it is certainly not a note of alarm. This domestic bird might therefore be considered a nocturnal songster, though we do not hear him at evening twilight. The cock sings his matins, but not his vespers. He crows at the earliest dawn and at midnight when he is wakened by the light of the moon, and by artificial light. Many birds are accustomed to prolong their notes after sunset to a late hour, and become silent only to begin anew at the earliest daybreak. But the habit of singing in the night is peculiar to a small number of birds, and the cause of it is a curious subject of inquiry.

By what means are they qualified to endure such extreme watchfulness, — singing and providing for their offspring during the day, then becoming wakeful and musical during the night? Why do they take pleasure in singing when no one will come in answer to their call? Have they their worship like religious beings; and are their midnight lays but the fervent outpouring of their devotions? Do they rejoice like the clouds in the presence of the moon, hailing her beams as a pleasant relief from the darkness that has surrounded them? Or, in the silence of the night, are their songs but responses to the sounds of the trees, when they bow their heads and shake their rustling leaves to the wind? When they listen to the streamlet that makes audible melody in the hush of night, do they not answer to it from their leafy perch?

And when the moth flies hummingly through the recesses of the wood, and the beetle winds his horn, what are the notes of the birds but cheerful counterparts to those sounds that break sweetly upon the quiet of their slumbers?

Wilson remarks that the hunters in the Southern States, when setting out on an excursion by night, as soon as they hear the mocking-bird, know that the moon is rising. He quotes a writer who supposes that it may be fear that operates upon the birds when they perceive the owls flitting among the trees, and that they sing as a timid person whistles in a lonely place to quiet their fears. But if such be the case, Nature has implanted in them an instinct that might lead to their destruction. Fear would instinctively prompt them to be quiet, if they heard the stirring of owls; for this feeling is not expressed by musical notes, but by notes of alarm, or by silence. The moonlight may be the most frequent exciting cause of nocturnal singing; but it is not true that birds always wait for the rising of the moon; and if it were so, the question still occurs, why a few species only should be thus affected.

Since philosophy cannot explain this instinct, let fancy come to our aid, as when men vainly seek from reason an explanation of the mysteries of religion they humbly submit to the guidance of faith. With fancy for our interpreter we may suppose that Nature has adapted the works of creation to our moral as well as our physical wants; and while she has instituted the night as a time of general rest, she has provided means that shall soften the gloomy effects of darkness. The birds, which are the harbingers of all rural delights, are hence made to sing during twilight; and when they cease, the nocturnal songsters become vocal, bearing pleasant sensations to the sleepless, and by their lulling melodies prepare us to be keenly susceptible to all agreeable emotions.

# CLOUDS.

THE sky would present very little in the daytime to charm the sight or interest the mind if it were destitute of clouds. From these proceed all the beautiful tints of sunrise and sunset, the rainbow, and the various configurations that deck the arches of the firmament. The different forms and colors they assume in their progress through the atmosphere, and their ever-varying positions and combinations, are capable of awaking the most agreeable emotions of beauty and sublimity. It is not often that the same object causes these two different emotions. But when the western clouds, piled in glittering arches one above another, and widening as they recede from the great source of light, display their several gradations of hues, from the outermost arch successively of violet, purple, crimson, vermilion, and orange; until our eyes are dazzled by the radiance that beams from the throne of day, the mind is affected with a sensation of beauty, accompanied by the most cheerful exaltation.

The great painters have delighted in the representation of clouds, knowing that every landscape may be improved by their celestial forms and tints, and that a scene representing any passion or situation may be heightened by such accompaniments, harmonizing with the cheerfulness or sadness, with the lowliness or magnificence of the subject. Poets have ever been mindful of the same effects; and the Hebrew prophets have exalted the sublimity of their descriptions and increased the efficacy of their prophecies and their admonitions by employing imagery derived from these appearances, adapted to illustrate their sacred themes. Hence Jehovah, who set his bow in a cloud as the token of a covenant between heaven and earth, is represented as making clouds his chariot and pavilion when ascending into heaven, or when descending on earth to speak to the messengers of his will.

Every scene in the universe is the cause, when we behold it, of a peculiar and specific sensation. Our emotions are as infinite as our thoughts, and Nature has provided an infinite variety of scenes to harmonize with all, that no existing susceptibility to pleasure shall be lost for the want of something external to act upon it, and render it a source of happiness. There are beams in the countenance of morn and even that irradiate into our souls a feeling of serene delight; and it is no marvel that Nature should seem, as the poets have described her, to smile upon us in the sunshine that sparkles in the morning dews and gilds the evening sky, or in the moonlight that reveals to us a new firmament of wonders among the silvery clouds of night. The forms and tints of clouds produce effects upon the mind that vary with the hour of the day. In the morning there is a feeling of hopefulness attending the spectacle of the constantly increasing splendor of the clouds, beginning with the dark purple tints of dawn, and widening with beautiful radiating undulations through their whole succession of hues into perfect day. As we are prepared by the buoyant feelings that come from the spectacle of dawn to enter with a glad heart upon the duties of the day, we are equally inspired by the spectacle of sunset with a sentiment of tranquillity that prepares us for healthful repose.

It is not difficult to understand that if the sun rose clearly into the blue heavens without any changes except from darkness to light, through all the degrees of twilight, the charms of the morning would be greatly diminished.

But Nature, that all hearts might be enamored of the morn, has wreathed her temples with dappled crimson, and animated her countenance with those milder glories that so well become the fair daughter of the dawn and the gentle mother of dews. In ancient fable, Aurora is a beautiful nymph who blushes when she first enters into the presence of Day, and the clouds are the fabric with which she veils her features at his approach. But a young person of sensibility needs no such allegory to inspire him with a sense of the incomparable beauty and grandeur of the orient at break of day. It is associated with some of the happiest moments of his life; and the exhilarated feelings with which we look upon the dayspring in the east are probably one cause of the tonic and healthful influence of early rising.

The forms of clouds are not less beautiful or expressive than their colors. While their outlines are sufficiently definite for picturesque effects, they often assume a great uniformity in their aggregations. The frostwork on our window-panes on cold winter mornings exhibits no greater variety of figures than that assumed by the clouds in their distribution over the heavens. Beginning in the form of vapor that rolls its fleecy masses slowly over the plain, resembling at a distance sometimes a smooth sheet of water, and at other times a drifted snow-bank, the cloud divides itself as it ascends, into globular heaps that reflect the sunlight from a thousand silvery domes. These, after gradually dissolving, reappear in a host of finely mottled images, resembling the scales of a fish, then marshal themselves into undulating rows like the waves of the sea, and are lastly metamorphosed into a thin gauzy fabric, like crumpled muslin, or in a long drapery of hair-like fringe, overspreading the higher regions of the atmosphere.

These different forms of cloud are elevated according to

the fineness of their texture and organization, the finer and more complicated fabrics occupying the space above the next in degree. We often observe three layers of cloud separated by sufficient space to receive all the different hues of sunset at the same moment. While the feather clouds that occupy the greatest elevation are burnished with a dazzling radiance, the middle layers of dappled cloud will be tipped with crimson, while the violet and indigo hues are seen in the dense unorganized mass that is spread out below. It may be remarked, both of the forms and the hues of clouds, that nature permits no harsh contrasts or sudden transitions. The different hues are laid softly one above another, melting into each other like those in the plumage of a dove. You can never see where one hue terminates and another commences. It is the same in a less degree with their forms, that never for two minutes in succession remain unaltered. They exhibit a pleasing irregularity, and are almost destitute of outlines, so that the imagination is left to carve out of their obscure figures and arrangements aerial landscapes, bright sunny valleys, and rolling plains, with villages surrounded by turrets and the pinnacles of mountains.

The imagination is always stimulated by a certain degree of obscurity in the objects of sight and sound as well as of thought. The sublime passages of the poets are often obscure, suggestive of something that produces a well-defined emotion, but no distinct image to the understanding. It is this quality that gives their power to certain remarkable passages in the Hebrew prophets. In a terrestrial landscape, when viewed by daylight, the outlines of objects, except at a distance, are so distinct that we can see and easily describe their forms and character. Distant objects have a dimness of outline and a misty obscurity which are favorable to an expression of sublimity. In the darkness of night the forms of trees display the indefinite shapes of clouds, and the imagination is free to indulge its caprices, while, as we pass by them in a journey or a ramble, our eyes are watching their apparent motions and changes of form.

By no other scenes in nature is the imagination so powerfully excited as by these celestial phenomena, whether we imagine the gates of heaven to be opened beneath the triumphal arches of sunset, or watch for the passing of the gloomy precursors of evil days in the dark irregular masses that deform the skies before a storm. The picturesque effects of clouds are in great measure attributable to the dubious character of their configurations, giving rise to peculiar fancies and awaking sentiments that spring only from the loftiest images of poetry. The shadows of passing clouds, as they fall upon the earth after moving rapidly with the wind, add greatly to their expression. Above all, do their motions contribute to the beauty of landscape, when, through some opening in their dense masses, while the greater part of the prospect is enveloped in shade, the sun pours a stream of glory upon a distant grove, village, or range of hills.

As the most delightful views of ocean are attained when a small part of it is seen through a green recess in a wood, for the same cause the blue sky is never so beautiful as when seen through the openings in the clouds. The emotion produced by any scene is the more intense when the greater part of the object that causes it is hidden, leaving room for the entrance of pleasant images into the mind. Clouds are peculiarly suggestive on account of the ambiguity of their shapes and their constant changes. Nothing, indeed, in nature so closely resembles the mysterious operations of thought, ever ceaseless in their motions and ever varying in their combinations, — now passing from a shapeless heap into a finely marshalled band; then dissolving into the pellucid atmosphere as a series of thoughts will pass away from our memory; then slowly forming themselves again and recombining in a still more beautiful and dazzling congeries in another part of the sky; now gloomy, changeable, and formless, then assuming a definite shape and glowing with the most lovely beams of light and beauty; lastly fading into darkness when the sun departs, as the mind for a short period is obliterated in sleep.

It is remarkable that in the evening, after the hues of sunset have faded to a certain point, the clouds are sometimes reilluminated before darkness comes on. Before the sun declines, the clouds are grayish tipped with silver. As he recedes, the gray portion becomes brown or auburn, and the silvery edges of a yellow or golden hue. While the auburn is resolved into purple, the yellows deepen into vermilion and orange. Every tint is constantly changing into a deeper one, until the sky is decorated with every imaginable tint except green and blue. When these colors have attained their greatest splendor, they gradually fade until the mass of each cloud has turned to a dull iron-gray, and every beautiful tint has vanished. We might then suppose that all this scene of glory had faded. After a few minutes, however, the clouds begin once more to brighten; the whole scene is gradually reilluminated, and passes through another equally regular gradation of more sombre tints, consisting of olive, lilac, and bronze, and their intermediate shades. The second illumination is neither so bright nor so beautiful as the first. But I have known the light that was shed upon the earth to be sensibly increased for a few moments by this second gradation of hues, without any diminution of the mass of cloud.

Men of the world may praise the effects of certain medical excitants that serve, by benumbing the outward senses, to exalt the soul into reveries of bliss and untried exercises of thought. But the only divine exhilaration proceeds from contemplating the beautiful and sublime scenes of nature as beheld on the face of the earth and the sky. It is under this vast canopy of celestial splendors, more than in any other situation, that the faculties may become inspired without madness and exalted without subsequent depression. I never believe so much in the divinity of nature as when, at sunset, I look through a long vista of luminous clouds far down into that mystic region of light in which we are fain to imagine are deposited the secrets of the universe. The blue heavens are the page whereon nature has revealed some pleasant intimations of the mysteries of a more spiritual existence; and no charming vision of heaven and immortality ever entered the human soul but the Deity responded to it upon the firmament in letters of gold, ruby, and sapphire.

# SOUNDS FROM ANIMATE NATURE.

A TREATISE on the beauties of nature would be very imperfectly accomplished if nothing were written of sounds. The hearing is indeed the most intellectual of our senses, though from the sight we undoubtedly derive the most pleasure. Hearing is also more intimately connected with the imagination than any other sense; and a few words of speech or a few notes of music may produce the most vivid emotion or awaken the most ardent passion. At all seasons and in all places the sounds no less than the visible things of nature affect us with pleasure or with pain. Everywhere does the song of a bird or the note of an insect, the cry of an animal or other sound from the animate world, come to the ear with messages of the past, conveying to the mind some joyful or plaintive remembrance.

Sounds are the medium through which many ideas as well as sensations are communicated to us by nature; and we cannot say how large a proportion of those which seem to rise spontaneously in the mind are suggested by some animal, through its cries of joy or complaint. There is hardly a rational being who is not alive to these suggestions, varying with his habits of life, especially those of his early years. Some persons do not purposely listen to the voices of insects, and seem almost unaware of the existence of these sounds. Yet even these apathetic persons are unconsciously affected by them. We attend so little to the subjects of our consciousness that we can seldom trace to their source any of our most ordinary emotions. We see without conscious observation and hear without conscious attention, so that when we are suddenly deprived of these sights and sounds we feel that there is a blank in our enjoyments, which can be filled only by those charming objects that never before received our thought or attention. How many bright things have faded on our mind, and how many sweet sounds have died on the ear before we were hardly aware of their existence!

If we hearken attentively to the miscellaneous sounds that come to our ears from the outer world, we shall perceive that some of them are cheerful and exhilarating, others are melancholy and depressing. Of the first are chiefly the songs of birds, the noise of poultry, the chirping of insects; indeed, the greater part of the sounds of animate nature. The second class comes chiefly from inanimate things, as the whistling of winds, the murmur of gentle gales, the roar of storms, the rush of falling water, and the ebbing and flowing of tides. All these are of a plaintive character, sometimes gloomy and sad, at other times merely soothing and tranquillizing. They all produce more or less of what physicians call a sedative effect. These two classes of sounds are often inseparably blended, inasmuch as some of the voices of birds, insects, and other creatures are melancholy, and some of the sounds of winds and waters are cheerful.

I shall treat of these different sounds chiefly as they affect the mind and sensibility; of the poetry rather than the science of these phenomena. My object is to point out one remarkable source of our agreeable sensations as derived from nature, and to show in what manner we may cause them to contribute to our pleasure. I am persuaded that one important means of deriving pleasure from any object is to direct our attention to it; and if this be not an indulgence that is liable to increase to a vicious extent, our happiness will be improved by our devotion to it. By studying the various sounds of nature and by habitually giving our attention to them, we become more and more sensitive to their influence and capable of hearing music to which others are deaf.

Cheerful sounds come chiefly from animated things; and from this we may infer that the mass of living creatures, in spite of the evils to which they are exposed and the pains they suffer, are happy. The chirping of insects denotes their happiness. No man goes out in the autumn and listens to the din of crickets and grasshoppers among the green herbs, and regards it as a melancholy sound. To all ears these notes express the joy of the creatures that utter them. Those doleful moralists who look upon everything as born to woe are greatly deluded; else why do not the voices of the sufferers give utterance to their pangs? Why, instead of uttering what seem like songs of praise, do they not cry out in doleful strains that would excite our pity? The greater part of the life of every creature is filled with agreeable sensations.

The fly, the gnat, the beetle, and the moth, though each makes a hum that awakens many pleasing thoughts and images, are not to be ranked among singing insects. Among the latter are crickets and locusts and grasshoppers, which are appointed by nature to take up their little lyre and drum after the birds have laid aside their more melodious pipe and flute. Their musical apparatus is placed outside of their bodies, and as they have no lungs, the air is obtained by a peculiar inflation of their chests. Hence the musical appendages of insects are constructed like reed instruments or jews'-harps. The grasshoppers in all ages have been noted as musical performers, and in certain ancient vignettes are frequently represented as playing on the harp.

Each species of insect has a peculiar modulation of his

notes. The common green grasshopper, that during the months of August and September fills the whole atmosphere with its din, abides chiefly in the lowland meadows which are covered with the native grasses. This grasshopper modulates its notes like the cackling of a hen, uttering several chirps in rapid succession and following them with a loud spinning sound that seems to be the conclusion of the strain. These notes are continued incessantly, from the time when the sun is high enough to have dried the dews until dewfall in the evening. The performers are delighted with the sunshine, and sing but little on cloudy days, even when the air is dry and warm.

SONG OF THE DIURNAL GREEN GRASSHOPPER.

There is another grasshopper with short wings that makes a kind of grating sound difficult to be heard, by scraping its legs, that serve for bows, upon its sides, that represent as it were the strings of a viol. If we go into the whortleberry pastures we hear still another species, that makes a continued trilling like the note of a hairbird. In some places this species sings very loudly, and continues half a minute or more without rest. Its notes are not so agreeable as those which are more rapidly intermittent.

There is a species of locust, seldom heard until midsummer, and then only in very warm weather. His note is a pleasant reminder of sultry summer noondays, of languishing heat and refreshing shade. The insect begins low, usually high up in the trees, and increases in loudness until it is almost deafening, and then gradually dies away into silence. The most skilful musician could not surpass his crescendo and diminuendo. It has a peculiar vibrating sound that seems to me highly musical and expressive. The insect that produces this note is a grotesque-looking creature, resembling about equally a grasshopper and a humblebee.

The black crickets and their familiar chirping are well known to everybody. An insect of this family is celebrated in English poetry as the "cricket on the hearth." Those of the American species are seldom found in our dwelling-houses; but they are all around our door-steps and by the wayside, under every dry fence and in every sandy hill. They chirp all day and some part of the night, and more or less in all kinds of weather. They begin their songs before the grasshoppers are heard, and continue them to a later period in the autumn, not ceasing until the hard frosts have driven them into their retreats and lulled them into a torpid sleep.

The note of the katydid, which is a mere drumming sound, is not musical. In American literature no insect has become so widely celebrated, on account of a fancied resemblance to the word "katydid." To my ear a chorus of these minute drummers, all uttering in concert their peculiar notes, seems more like the hammering of a thousand little smiths in some busy hamlet of insects. There is no melody in these sounds, and they are accordingly less suggestive than those of the green nocturnal grasshopper, that is heard at the same hour in similar situations.

The nocturnal grasshoppers, called August pipers, or Cicadas, begin their chirping about the middle of July, but are not in full song until August. These are the true nightingales of insects, and the species that seems to me the most worthy of being consecrated to poetry. There is a singular plaintiveness in their low monotonous notes, which are the charm of our late summer evenings. There are but few persons who are not affected by these sounds with a sensation of subdued but cheerful melancholy. This effect does not seem to be caused by association so much as by their peculiar cadence. The notes of these nocturnal pipers on very warm evenings are in unison and accurately timed, as if they were singing in concert. It is worthy of notice that they always vary their keynote according to the temperature of the atmosphere. They are evidently dependent on a summer heat for their vivacity, and become sluggish and torpid as the thermometer sinks below a certain point. When the temperature is high they keep good time, singing shrilly and rapidly. As it sinks they take a lower key and do not keep time together. When the thermometer is not above sixty, their notes are very low, and there are but few performers.

| Height of Thermometer. | Keynote of the Insects. |
|---|---|
| 80° | F natural, perfect time and tune. |
| 75° | E flat, " " " |
| 70° | D, " " " |
| 65° | C, imperfect time and tune. |
| 60° | B flat, " " " |
| 55° | A, keynote hardly to be detected, many out of time and tune. |
| 50° | G, a few individuals only, singing slowly and feebly. |

## OCTOBER.

THE cool and temperate breezes that prevail at this time almost constantly from the west, attended with a clear sky, announce the brilliant month of October with a climate that alternately chills the frame with frosty vapors by night and enlivens the heart with beauty and sunshine by day. At sunrise the villagers are gathered round their fires shivering with cold; the chirping insects also have crept into their shelters and are silent. But ere the sun has gained half his meridian height the villagers have forsaken their fires, and are busy in the orchards beneath the glowing sunshine; and the insects, aroused from their torpor and warmed into new life, are again chirping as merrily as in August, and multitudes that could hardly creep with torpor in the morning are now darting and spinning in the grassy meadows.

There are occasional dull and cloudy days in October, the dreary precursors of approaching winter; but they are generally bright and clear, and unequalled by those of any other month in salubrity. There are no sleeping mists drawn over the skies to obscure the transparency of the atmosphere; but far as the eye can reach, the distant hills lift up their heads with a clear, unclouded outline, and the blue arch of heaven preserves its deep azure down almost to the horizon. In the mornings of such days a white fleecy cloud is settled upon the streams and lowlands, in which the early sunbeams are refracted with all the myriad hues of dawn, forming halos and imperfect rainbows that seem to be pictured on a groundwork of drifted snow. By this vapor, nearly motionless at sunrise, we may trace the winding course of the small rivers far along through the distant prospect. But the sun quickly dissipates this fleecy cloud. As the winds float it slowly and gracefully over the plains, it melts into transparency; and ere the sun has gained ten degrees in his orbit, the last feathery fragment has vanished and left him in the clear blue firmament without one shadow to tarnish his glory.

October is the most brilliant of the months, unsurpassed in the clearness of its skies and in the wonderful variety of tints that are sprinkled over all vegetation. He who has an eye for beautiful colors must ever admire the scenery of this last month of foliage and flowers. As Nature loses the delicacy of her charms, she is more lavish of the gaudy decorations with which she embroiders her apparel. While she appears before us in her living attire, from spring to autumn she is constantly changing her vesture with each passing month. The flowers that spangle the green turf or wreathe themselves upon the trees and vines, and the herbage with all its various shades of verdure, constitute, with their successive changes, her spring and summer adornment; but ere the fall of the leaf she makes herself garlands of the ripened foliage, and crowns the brows of her mountains and the bosoms of her groves with the most beautiful array.

Though the present is a melancholy time of the year, we are preserved from cheerless reflections by the brightness of the sunshine and the interminable beauty of the landscape. The sky in clear weather is of the deepest blue; and the ocean and the lakes, slightly ruffled by the October winds, which are seldom tranquil, have a peculiar depth of coloring, unwitnessed when their surface is calm. Diverted by the unusual charms of Nature, while we look with a mournful heart upon the graves of the flowers, we turn our eyes upward and around us, where the woods are glowing like a wilderness of roses, and forget in our ravishment the beautiful things we have lost. As the flowers wither and vanish from our sight, their colors seem to revive in the foliage of the trees, as if each dying blossom had bequeathed its beauty to the forest boughs, that had protected it during the year. The trees are one by one putting aside their vestures of green and slowly assuming their new robes of many hues. From the beginning to the end of the month the landscape suffers a complete metamorphosis; and October may be said to represent in the successive changes of its aspect all the floral beauty of spring and summer.

Unaffected by the late frosts, the grass is still green from the valleys to the hill-tops, and many a flower is still smiling upon us as if there were no winter in the year. Many fair ones still linger in their cheerful but faded bowers, the emblems of contentment, seeming perfectly happy if they can but greet a few beams of sunshine to temper the frosty gales. In wet places I still behold the lovely neottia with its small white plumes arranged in a spiral line about their stems, and giving out the delicate incense of a lily. The purple gerardia, too, has not yet forsaken us, and the gentians will wait till another month before they wholly leave our borders.

If we quit the fields we find in the gardens a profusion of lovely exotics. Dahlias and fuchsias, and many other plants that were created to embellish other climes, are rewarding the hands that cherished them with their fairest forms and hues. All these are destined, not, like the flowers of our own clime, to live throughout their natural period, and then sink quietly into decay, but to be cut down by frosts in the very summer of their loveliness. Already are their leaves withered and blackened, while the native plants unseared by the frost, grow brighter and brighter with every new morning, until they are finally seared by the icy breath of November.

But to the forests we must look to behold the fairest spectacle of the season, now glowing with the infinitely varied and constantly multiplying tints of a summer sunset. The first changes appear in the low grounds, where vegetation is exposed to the earliest blights, and is prematurely ripened by the alternation of chill dews and sunshine. Often in the space of one night the leaves of the trees are metamorphosed into flowers, as if the dewdrops brought with them the hues of the beautiful clouds from which they fell. But Nature, while decorating some trees in one uniform color, scatters over the remainder a gentle sprinkling of every hue.

It is my delight during this month to ramble in the field and wood, to take note of these changes as they happen day by day. Each morning witnesses a new aspect in the face of Nature like each passing moment that attends the brightening and fading of the evening sky. The landscape we visited but a few days since is to-day like a different prospect, save in the arrangement of the grounds. Beauty has suddenly awoke upon the face of a dull and homely wood, and variety has sprung up in the midst of tiresome uniformity. There are patches of brightly tinted shrubbery that seem to have risen during the night from the bed of the earth where yesterday there was but a dull uniform green, and· when surrounded by the unfaded grasses, they resemble little flower-plats embosomed in verdure. As the month advances one tree after another partakes of this beautiful transformation. All the shades of red, yellow, and purple are resplendent from different species. It seems as if the departed flowers of summer had revisited the earth, and were wreathing their garlands around the brows of the woods and the mountains.

On every side of our walk various plats of herbage gleam upon our sight, each with some unmingled shade of some lovely hue; and every shrub and every leafy herb presents the appearance of a scattered variety of bouquets, wreaths, and floral embroidery. The farms in the lowlands display wide fields of intermingled orange and russet, and the shrubs of different colors that spring up among them in clumps and knolls add to the spectacle an endless variety of splendor. The creeping herbs and trailing vines, some begemmed with fruit, display the same variety of tinting, as if designed for wreaths to garland the gray rocks, and to yield a smile to the face of Nature that shall make glad the heart of the solitary rambler, who is ready to weep over the fair objects that have fled.

Day and night have at length about equally divided the light and the darkness. The time of the latter harvest is nearly past, and the winter fruits are mostly gathered into barns. The mornings and evenings are cold and cheerless, and the west-wind has grown harsh and uncomfortable. The bland weather of early autumn is rapidly gliding from our year. Night is continually encroaching upon the dominion of day. The white frosts already glitter in the arbors of the summer dews, and the cold north-wind is whistling rudely in the haunts of the sweet summer zephyrs. The scents of fading leaves and of the ripened harvest have driven out the delicate incense of the flowers whose fragrant offerings have all ascended to heaven. Dark threatening clouds occasionally frown upon us as they gather for a few hours about the horizon, the melancholy omens of the coming of winter. But there is pleasantness still in a rural excursion, and when the cold mists of dawn have passed away and the hoar-frost has melted in the warm sunshine, it is my delight to go out into the field to take note of the last beautiful things of summer that linger on the threshold of autumn.

# CHANGES IN THE HABITS OF BIRDS.

BIRDS acquire new habits as certain changes take place upon the surface of the country that create a necessity for using different modes of sheltering and protecting their young. Singing-birds frequent in greatest numbers our half-cultivated lands and the woods adjoining them. It may therefore be inferred that as the country grows older and is more extensively cleared and cultivated, the numbers of our songsters will increase, and it is not im-. probable that their vocal powers may improve. It may be true that for many years after the first settlement of this country there were but few singing-birds and that they have multiplied with the cultivation of the soil. At that time, though the same species existed here and were musical, their numbers were so small that they were not universally heard. Hence early travellers were led to believe that American birds were generally silent.

By a little observation we should soon be convinced that the primitive forest contains but few songsters. There you find crows, jays, woodpeckers, and other noisy birds in great numbers ; and you occasionally hear the notes of the sylvias and solitary thrushes. But not until you are in the vicinity of farms and other cultivated lands are your ears saluted by a full band of feathered musicians. The bobolinks are not seen in a forest, and are unfrequent in the wild pastures or meadows which were their primitive resorts. At the present day they have left their early habitats, and seek the cultivated grass-lands, that afford them a more abundant supply of insect-food, with which they feed their young. They build upon the ground in the grass, and their nests are exposed in great numbers by the scythe of the mower, if he begins haymaking early in the season.

These birds, as well as robins, before America was settled by the Europeans, and when the greater part of the country was a wilderness, must have been comparatively few. Though the bobolink consumes great quantities of rice after the young are fledged and the whole family have departed, it is not the rice-fields which have made its species more numerous, but the increased abundance of insect food in the North, where they breed, — an increase consequent upon the increased amount of tillage. The robins are dependent entirely upon insect food, and must have multiplied in greater proportion than the bobolinks. There are probably thousands of both species at the present day to as many hundreds that existed at the discovery of America. Many other small birds, such as the song-sparrow and the linnet, have increased nearly in the same ratio with the progress of agriculture and the settlement of the country.

Domestication blunts the original instincts of animals and renders birds partially indifferent to colors. It changes their plumage as well as their instincts. In proportion to the length of time any species has been domesticated, it is unsafe to depend on the correctness of our observation of their instincts with respect to colors. All the gallinaceous birds, except the common hen, lay speckled eggs. It is probable that during the thousands of ages since the latter was domesticated her eggs have lost their original marking and have become white. As great a change has happened in their plumage, while the more recently domesticated birds, like the turkey and guinea-hen, retain more nearly their original markings. After domestication birds no longer require to be protected from the sight of their enemies by the hues of their plumage. Their natural predisposition to be marked only by a certain combination of hues is weakened. Being entirely in the power and under the protection of man, color is of no service to them, as in their natural and wild state.

Mr. S. P. Fowler communicated to the Essex Institute an essay containing some important facts concerning the changes in the habits of some of our own birds. He says: " The Baltimore oriole still constructs her nest after the old pattern, but has learned to weave it with materials furnished by civilization. I have a whole nest of this kind, made wholly from materials swept out of a milliner's shop, woven and interlaced with ribbons and laces, including a threaded needle." He has noticed for several years a change in the habits of our crow-blackbirds, and thinks they are becoming domesticated, like the rooks of England. This change, in his opinion, has been produced by planting the white pine in cultivated grounds; for wherever a group of pines has attained the height of thirty feet, they are visited by these birds for breeding, even in proximity to our populous villages. He states that the purple finches have followed the evergreen trees that have been planted in our enclosures, though a few years since they were to be seen chiefly in our cedar groves. They have grown more numerous, and breed in his grounds on the branches of the spruce, feeding early in the season upon the flower-buds of the elm or upon those of the pear-tree.

From the same communication I gather the following facts, slightly abridging his statements. He remarks that the swallows have suffered more changes than any other birds of our vicinity. The barn-swallows long since left their ancient breeding-places, the overhanging cliffs of rocks, and have sought buildings erected by man; the chimney-swallow has deserted the hollow sycamore for some deserted chimney; and the cliff-swallow has left the shelving rock to seek shelter under the eaves of our roofs. The purple martin and white-bellied swallow have left the wilderness to find a home in our villages. The purple martins, during the last fifty years, have gradually diminished in Eastern Massachusetts. He thinks it equally certain that the barn-swallows are growing less numerous, and attributes their diminution to our modern tight barns. Chimney-swallows, on the other hand, have become more numerous. The opening of the Pacific Railroad, he thinks, will cause both plants and birds to follow its track.

CALIFORNI

# BIRDS OF THE MOOR.

### THE AMERICAN WOODCOCK.

THE American Woodcock is a more interesting bird than we should suppose from his general appearance and physiognomy. He is mainly nocturnal in his habits, and his ways are very singular and worthy of study. He obtains his food by scratching up the leaves and rubbish that lie upon the surface of the ground in damp and wooded places, and by boring into the earth for worms. He remains concealed in the wood during the day, and comes out to feed at twilight, choosing the open ploughed land where worms are abundant. Yet it is probable that in the shade of the wood he is more or less busy among the leaves in the daytime.

The Woodcock does not usually venture abroad in the open day, unless he be disturbed and driven from his retreat. He makes his first appearance here early in April, and at this time we may observe that soaring habit which renders him one of the picturesque objects of nature. This soaring takes place soon after sunset, continues during twilight, and is repeated at a corresponding hour in the morning. If you listen at these times near the place of his resort, he will soon reveal himself by a lively peep, frequently uttered from the ground. While repeating this note he may be seen strutting about like a Turkey-cock, with fantastic jerkings of the tail and a frequent turning of the head; and his mate is, I believe, at this time not far off. Suddenly he springs upward, and with a wide circular sweep, uttering at the same time a rapid whistling note, he rises in a spiral course to a great height in the air. At the summit of his ascent, he hovers about with irregular motions, chirping a medley of broken notes, like imperfect warbling. This continues about ten or fifteen seconds, when it ceases and he descends rapidly to the ground. We seldom hear him in his descent, but receive the first intimation of it by the repetition of his *peep*, like the sound produced by those minute wooden trumpets sold at the German toy-shops.

No person could watch this playful flight of the Woodcock without interest; and it is remarkable that a bird with short wings and difficult flight should be capable of mounting to so great an altitude. It affords me a vivid conception of the pleasure with which I should witness the soaring and singing of the Skylark, known to us only by description. I have but to imagine the chirruping of the Woodcock to be a melodious series of notes to feel that I am listening to the bird which has been so familiarized to us by English poetry, that in our early days we often watch for his greeting on a summer sunrise. It is with sadness we first learn that the Skylark is not an inhabitant of the New World; and our mornings and evenings seem divested of a great part of their charm by their want of this lyric accompaniment.

There are other sounds connected with the flight of the Woodcock that increase his importance as an actor in the great melodrama of Nature. When we stroll away at dusk from the noise of the town, to a spot where the stillness permits us to hear distinctly all those faint sounds which are turned by the silence of night into music, we may hear at frequent intervals the hum produced by the irregular flight of the Woodcock as he passes over short distances near the wood. It is like the sound of the wings of Doves, or like that produced by the rapid whisking of a slender rod through the air. There is a plaintive feeling of mystery attached to these musical flights that yields a savor of romance to the quiet voluptuousness of a summer evening.

On such occasions, if we are in a moralizing mood, we are agreeably impressed with the truth of the maxim that the secret of happiness consists in keeping alive our susceptibilities by frugal indulgence, and by avoiding an excess of pleasures that pall in proportion to their abundance. The stillness and darkness of a quiet night produce this quickening effect upon our minds. Our susceptibility is then awakened to such a degree that slight sounds and faintly discernible lights convey to us an amount of pleasure that is seldom felt in the daytime from influences even of a more inspiring character. Thus the player in an orchestra can enjoy such music only as would deafen common ears by its crash of sounds in which they can perceive no connection or harmony; while the simple rustic listens to the rude notes of a flageolet in the hands of a clown with feelings of ineffable delight. To the seekers after luxurious and exciting pleasures, Nature, if they could but understand her language, would say, "Except ye become as this simple rustic, ye cannot enter into my paradise."

THE SNIPE.

The Snipe has the nocturnal habits of the Woodcock, and is common in New England in the spring and autumn, but does not often breed here. It has the same habits of feeding as the Woodcock, and the same way of soaring into the air during morning and evening twilight, when he performs a sort of musical medley, which Audubon has described in the following passage: "The birds are met with in the meadows and low grounds, and by being on the spot before sunrise, you may see both male and female mount high in a spiral manner, now with continuous beats of the wings, now in short sailings, until more than a hundred yards high, when they whirl round each other with extreme velocity, and dance as it were to their own music; for at this juncture, and during the space of four or five minutes, you hear rolling notes mingled together, each more or less distinct, perhaps, according to the state of the atmosphere. The sounds produced are extremely pleasing, though they fall faintly on the ear. I know not how to describe them; but I am well assured that they are not produced simply by the beatings of their wings, as at this time the wings are not flapped, but are used in sailing swiftly in a circle, not many feet in diameter. A person might cause a sound somewhat similar, by blowing rapidly and alternately from one end to another across a set of small pipes consisting of two or three modulations. This performance is kept up till incubation terminates; but I have never observed it at any other period." In this respect the Snipe differs from the Woodcock, whose nocturnal flights I have not witnessed except in April and perhaps the early part of May. The time occupied by the Woodcock in the air is never more, I am confident, than fifteen seconds, and the notes uttered by him while poised at the summit of his ascent sound exactly like *chip, chip, chip, chip, chip, chip,* about as rapidly as we might utter them in a loud whisper.

### THE VIRGINIA RAIL.

The shyness and timidity of the Virginia Rail, and the quickness of its movements, its peculiar graceful attitudes, and the rare occasions on which we can obtain sight of one, combine to render this bird highly interesting. It is so seldom seen on account of its habit of concealment during the day and of feeding at evening and morning twilight, that many persons have never met with it. It is in fact quite a common bird, and breeds in the thickets in the immediate vicinity of our rivers and ponds. I have seen numbers of this species in the meadows surrounding Fresh Pond in Cambridge when hunting for aquatic plants and flowers; but I have not discovered their nests. Samuels says the eggs, which are from six to ten in number, are of a deep buff color, and that their nest "is nothing but a pile of weeds or grass which it arranges in a compact manner, and hollows to the depth perhaps of an inch or an inch and a half."

This is a very pretty species. The upper parts are brown, striped with deeper shades of the same color; the feathers on the breast are of a bright brown deepening into red; the wings black and chestnut with some white lines. It resembles somewhat a miniature hen with long legs and short tail, and is very nimble in its movements. This species is most commonly found in those fresh meadows into which the salt water extends or those salt marshes which are pervaded by a stream of fresh water. They feed more on worms and insects than upon seeds and grain, though they do not refuse a granivorous diet.

## THE CLAPPER RAIL.

I have so seldom seen the Clapper Rail, though I have many times heard its clattering notes, that I have nothing to say of it from my own observation. But as it is not unfrequent on the New England coast, it seems a fit subject to be introduced in my descriptions of picturesque birds. I shall, therefore, in this case deviate from my general practice of writing from my own experience, and insert in this place a brief abstract of an essay on

"The Clapper Rail," by Dr. E. Coues, published in the "American Naturalist," Vol. III. pp. 600–607.

The Clapper Rail, or Salt-water Marsh Hen, inhabits the marshes all along our coast, within reach of the tides, rarely, if ever, straying inward. It goes as far as Massachusetts, where it is rare; but is found abundantly in the Middle States, and in countless numbers on the coast of North Carolina, where it spends the whole year. The young birds while in their downy plumage are jet black, with a faint gloss of green, resembling newly hatched chickens. Rails live in the marshes, and are not very often seen except when they fly up.

The eggs of the Clapper Rail are of a pale buff or cream color. They are dotted or splashed with irregular spots of a dull purple or lilac color; and the number found in a nest is from six to nine. They raise two broods in a season, and some idea of the countless numbers of Rails in the marshes may be gained from the fact that baskets full of eggs are gathered by boys and brought to the Beaufort market.

The Rails' nests are sometimes floated away and destroyed by an unusual rise of the tide caused by a storm. A great tragedy of this kind happened at Fort Macon on the 22d of May, 1869, when the marsh, usually above water, was flooded, — only here and there a little knoll breaking the monotony of the water. There was a terrible commotion among the Rails at first, and the reeds resounded with their hoarse cries of terror. But as the waters advanced and inundated their houses the birds became silent again, as if in unspeakable misery. They wandered in listless dejection over beds of floating wrack, swam aimlessly over the water, or gathered stupefied in groups upon projecting knolls. Few of the old birds probably were drowned, but most of the young must have perished.

As if to guard against such an accident, the Rails generally build their nests around the margins of the marsh or in elevated spots, at about the usual high-water mark. The nest is always placed on the ground, in a bunch of reeds or tussock of grass or clump of little bushes. It is a flimsy structure made of dry grasses or reed-stalks broken in pieces and matted together, but not intertwined. Sometimes it is barely thick enough to keep the eggs from the wet.

The Rail, though not formed like a natatorial bird, swims very well for short distances. Dr. Coues has often seen it take to the water from choice, without necessity, and noticed that it swam buoyantly and with ease, like a coot. But the bird is a poor flyer, and it is surprising, therefore, that some of the family perform such extensive migrations. The Rails, in fact, are not distinguished either as flyers or swimmers. But as walkers they are unsurpassed; and have the power of making a remarkable compression of their body, that enables them to pass through close-set reeds. The bird indeed, when rapidly and slyly stealing through the brush, becomes literally as "thin as a rail."

Rails are among the most harmless and inoffensive of birds. But when wounded or caught, they make the best fight they can and show good spirit. In this case they use their sharp claws for a weapon rather than their slender bill. A colony of Rails goes far towards relieving a marsh of its monotony. Retiring and unfamiliar as they are, and seldom seen, considering their immense numbers, they have at times a very effective way of asserting themselves. Silent during a great part of the year, or at most only indulging in a spasmodic croak now and then, during the breeding-season they are perhaps the noisiest birds in the country. Let a gun be fired in the marsh, and like the reverberating echoes of the report a hundred cries come instantly from as many startled throats. The noise spreads on all sides, like ripples on the water at the plash of a stone, till it dies away in the distance. In the evening and morning particularly, the Rails seem perfectly reckless, and their jovial if unmusical notes resound till the very reeds seem to quake. Dr. Coues compares them to the French *claqueurs*. Unobtrusive, unrecognized except by a few, almost unknown to the uninitiated, the birds steadily and faithfully fulfil their allotted parts; like *claqueurs* they fill the pit, ready at a sign to applaud anything that may be going on in the drama of life before them.

### THE HERON.

No family of birds is possessed of more of those qualities which are especially regarded as picturesque than the Herons. This family comprehends a great many species, distinguished by their remarkable appearance both when flying aloft and when wading in their native swamps. They are generally seen in flocks, passing the day in sluggish inactivity, but called forth to action by hunger in the evening when they take their food. It is at the hour just after twilight that their peculiar cries are heard far aloft as they pass from their secluded day-haunts to their nocturnal feeding-places. Their flight deserves attention from their slow and solemn motion on the wing. Their flying attitude, however, is uncouth, with the neck bent backwards, their head resting against their shoulders, and their long legs stretched out behind them in the most awkward manner.

### THE BITTERN.

Among the Heron family we discover a few birds which, though not very well known, have ways that are singular and interesting. Goldsmith considered one of these worthy of introduction into his "Deserted Village" as contributing to the poetic sentiment of desolation. Thus, in his description of the grounds which were the ancient site of the village, we read:—

"Along its glades, a solitary guest,
The hollow-sounding Bittern guards its nest."

The American Bittern is a smaller bird than the one to which the poet alludes, but is probably a variety of the European species. It displays the same nocturnal habits, and has received at the South the name of *Dunkadoo*, from the resemblance of its common note to those syllables. This is a hollow-sounding noise, which would attract the attention of every listener. I have heard it by day in wooded swamps near ponds, and am at a loss to explain how so small a bird can produce so low and hollow a note. The common people of England have a notion that it thrusts its head into a hollow reed and uses it as a speaking-trumpet, and at times puts its head into the water and bubbles its notes in imitation of a bullfrog. The American Bittern utters another note resembling the sound produced by hammering upon a stake when driving it into the ground. Hence the name of *Stake-driver* applied to him in some parts of New England.

### THE QUA BIRD.

On a still evening in summer no sound is more common above our heads than the singular voice of the Qua Bird, as he passes in slow and solemn flight from his retreats where he passes the day to his feeding-places upon the sea-shore. His note is like the syllable *quaw* suddenly pronounced. If it were prolonged it might resemble the cawing of a Crow. This note is very frequently repeated, though one note by the same bird is never immediately succeeded by another. The birds of this species are social in their habits, and the woods in which they assemble are called heronries. During the breeding-season they are extremely noisy, uttering the most uncouth and unmusical sounds that can be imagined.

### THE CRANE, OR BLUE HERON.

The Crane is a very attractive bird; but the only individuals of the species I have seen enough to study their ways and manners were tamed. There is a sort of majesty in their appearance which I could not but admire. "During the day," says Samuels, "the Crane seems to prefer the solitudes of the forest for its retreat, as it is usually seen in the meadows only at early morning and in the latter part of the afternoon. It then, by the side of a ditch or a pond, is observed patiently watching for its prey. It remains standing motionless, until a fish or a frog presents itself, when with an unerring stroke with its beak, as quick as lightning it seizes, beats to pieces, and swallows it. This act is often repeated; and as the Heron varies this diet with meadow-mice, snakes, and insects, it certainly does not lead the life of misery and want that many writers ascribe to it."

This bird, like the Night Heron, breeds in communities. Samuels once visited with some attendants a heronry of this species in a deep swamp, intersected by a branch of the Androscoggin River. The swamp over which he had to pass was full of quagmires; and these he could hardly distinguish from the green turfy ground. It was only by wading through mud and water, sometimes nearly up to his waist, or by leaping from one fallen tree to another, through briers and brushwood, that he arrived beneath the trees which the birds occupied. These were dead hemlocks, without branches less than thirty feet from the ground, and could not be climbed. The nests, placed in the summits of the trees, were nearly flat, constructed of twigs and put together very loosely. It was on the 25th of June, and the young were about two thirds grown. He says the old birds flew over their heads uttering their hoarse, *husky*, and guttural cries. He observed, however, that they were careful to keep out of gunshot. The eggs, he says, are of a bluish-green color, and but one brood is reared in the season. The birds are very suspicious; they are constantly looking out for danger, and with their keen eyes, long neck, and fine sense of hearing, they immediately detect the approach of a gunner.

## SOUNDS FROM INANIMATE NATURE.

NATURE in every scene and situation has established sounds which are indicative of their character. The sounds we hear in the hollow dells among the mountains are unlike those of the open plains; and the echoes of the sea-shore repeat sounds never reverberated in the inland valleys. The murmuring of wind and the rustling of foliage, the gurgling of streams and the bubbling of fountains, come to our ears like the music of our early days, accompanied with many agreeable fancies. A stream rolling over a rough declivity, a fountain bubbling up from a subterranean hollow, give sounds suggestive of fragrant summer arbors, of cool retreats and all their delightful accompaniments.

The most agreeable expression from the noise of waters is their animation. They give life to the scenes around us, like the voices of birds and insects. In winter especially they make an agreeable interruption of the general stillness, and remind us that during the slumber of all visible things some hidden power is still guiding the operations of Nature. The rapids produced by a small stream flowing over some gentle declivity yield, perhaps, the most expressive sound of waters, save the distant roar of waves as they are dashed upon the sea-shore. The last, being intermittent, is preferable to the roar of a waterfall, which is tiresomely incessant. Nearly all the sounds made by water are agreeable, and cannot be multiplied without increasing the delightful influences of the place and the season.

Each season of the year has its peculiar melodies beside those proceeding from animated objects. In the opening of the year, when the leaves are tender and pliable, there is a mellowness in the sound of the breezes, as if they felt the voluptuous influence of spring. Nature then softens all the sounds from inanimate things, as if to avoid making any harsh discords with the anthem that issues from the woodlands, vocal with the songs of myriads of happy creatures. The echoes repeat less distinctly the multitudinous notes of birds, insects, and reptiles. To the echoes spring and summer are seasons of comparative rest, save when residing among the rocks of the desert or among the crags of the sea-shore. Here sitting invisibly in these retreats, they are ever responding to the melancholy sounds that are borne upon the waves as they sullenly recount the perils and accidents of the great deep.

But there are reverberations which are too refined and subtle to be distinguished as echoes. All creation, indeed, is a vast assemblage of musical instruments, whose chords vibrate to every sound in Nature. Every sound that peals over the landscape is in communication with millions of harps whose strings give out some response in harmony with the season and situation. As every ray of light coming from the farthest perceptible distance in the universe is repeated millions of times in various forms of beauty from dews and gems and flowers, — in the same manner do the sounds in the atmosphere vibrate from every spear of grass and every leaf of the forest, producing some unconscious pleasure.

After the frosts of autumn the winds become shriller as they pass over the naked reeds and rushes and through the leafless branches of the trees, and there is a familiar sadness in their murmurs, as they whirl among the dry rustling leaves. When the winter has arrived and enshrouded all the landscape in snow, the echoes venture out once more on the open plain, and repeat with unusual distinctness the various sounds from wood, village, and farm. During the winter they enjoy a long heyday of freedom; they hold a laughing revelry in the haunts of the dryad, and seem to rejoice as they sing together over the desolate appearance of Nature.

When the sun gains a few more degrees in his meridian height, and the snow begins to disappear under the fervor of his beams, then do the sounds from the dropping eaves and the clash of falling icicles from the boughs of the orchard-trees afford a pleasant sensation of the change; and the utterance of these vernal promises awakens all the delightful anticipations of birds and flowers. The moaning of winds has been plainly softened by the new season, and the summer zephyrs, that occasionally pay us a short visit from the south, and signalize their coming by the crimsoned dews at sunrise, loosen a thousand rills, that make lively music as they leap down the hill-sides into the valleys. Yet of all these sounds from inanimate Nature, there is not one but is hallowed by some glad or tender sentiment, of which it is suggestive, and we have but to yield our hearts to their influences to feel that for the ear as well as for the eye, Nature has provided an endless store of pleasure.

I believe the agreeable sounds from the inanimate world owe their principal effect to their power of gently exciting the sentiment of melancholy. The murmur of gentle gales among the trembling aspen-trees, the noise of the hurricane upon the sea-shore, the roar of distant waters, the sighing of wind as it flits by our windows and moans through the casement, have the power of exciting just enough of the sentiment of melancholy to produce an agreeable state of the mind. Along with the melancholy they excite there is something that tranquillizes the soul and exalts it above the mere pleasures of sense.

It is this power to produce the sentiment of melancholy that causes the sound of rain to afford pleasure. The pattering of rain upon our windows, but more especially upon the roof of the house under which we are sitting, is attended with a singular charm. There are few persons who do not recollect with a sense of delight some adventure in a shower, that obliged them on a journey to take shelter under a rustic roof by the wayside. The pleasure produced by the sight and sound of the rain under this retreat often comes more delightfully to our memory than all the sunshiny adventures of the day. But in order to be affected in the most agreeable manner by the sound of rain, it is necessary to be in company with those whom we love, or to feel an assurance that the objects of our care are within doors, and to be ignorant of any person's exposure to its violence.

During a thunder-storm the thunder is in many cases too terrific to allow us to feel a tranquil enjoyment of the occasion. There is no sound in Nature that is so pleasantly modified by distance. Some minutes before the thunder-storm there is a perfect stillness of the atmosphere which is fearfully ominous of the approaching tempest. It follows the first enshrouding of daylight in the clouds which are slowly gathering over our heads, as they come up from the western horizon. It is at such times that the sullen moan of the thunder, far down as it were below the belt of the hemisphere, is peculiarly solemn and impressive, and more productive of the emotion of sublimity than when the crash is heard directly over our heads. To be witness of a storm is pleasant when we are, and believe others to be, in a place of safety. Then do we listen with intense delight to the voice of winds and waters as they contend with the Demon of the storm, and the awful warring of the elements excites the most sublime sensations, unalloyed with any painful anxiety for the safety of a fellow-being.

Thunder is heard with different emotions when it proceeds from clouds which are moving towards us and when from those already settled down in the east, after the storm is past. The consciousness that the one indicates a rising storm renders forcibly suggestive the perils we are soon to encounter and increases our anxiety. When we are in the midst of the storm we feel the emotion of terror rather than that of sublimity. An uncomfortable amount of anxiety destroys that tranquillity of mind which is necessary for the full enjoyment of the sublime as well as the beautiful scenes of Naure.

It is pleasant after the terrors of the storm have ceased, when the blue sky in the west begins to appear in dim streaks through the misty and luminous atmosphere, to watch the lightnings from a window, as they play down the dark clouds in the eastern horizon, and to listen to the rumblings of the thunder as it begins loudly overhead, then dies away almost like the roaring of waves in a distant part of the heavens. Then do we contemplate the spectacle with a grateful sense of relief from the fears that lately agitated the mind, and surrender our souls to all the influences naturally awakened by a mingled scene of beauty and grandeur.

The emotion of sublimity is more powerfully excited by any circumstances that add mystery to a scene or to the sounds we may be contemplating. Hence any unknown sound that resembles that of an earthquake impresses the mind at once with a feeling of awe, however insignificant its origin. The booming of a cannon over a distance that renders its identity uncertain causes in the hearers a breathless attention, as to something ominous of danger. We may thus explain why all sounds are so suggestive in the stillness of night: the rustling of a zephyr as it glides half noiselessly through the trees; a few heavy drops of rain from a passing cloud, the signal of an approaching storm; the footfall of a solitary passenger on the road; the tinkling of a cowbell, heard now and then from a neighboring field, — all these are dependent on the stillness and darkness of the night for their influence on the mind.

It is evident that the charm of all these sounds is exalted by the imagination. A person who has not cultivated this faculty is deaf to a thousand pleasures from this source that form a considerable part of the happiness of a man of sensibility. Music has no advantage over other sounds save its greater power to act upon the imagination. To appreciate the charm of musical notes, or to perceive the beauty of an elegant house or of splendid tapestry, requires no mental culture. But to be susceptible of pleasure from what are commonly regarded as indifferent sounds is the meed only of those who have cherished the higher faculties and the better feelings of the soul. To such persons the world is full of suggestive sounds as well as suggestive sights, and not the whisper of a breeze or the murmur of a wave but is in unison with some chord in their memory or imagination.

# OLD ROADS.

I CANNOT say that I am an admirer of those tasteful operations which are commonly termed improvements, and seldom observe them without a feeling of regret. More of the beauty of landscape is destroyed every year by attempts to improve it, than by the ignorant or avaricious woodman who cuts down his trees for the railroad or the shipyard. There is a certain kind of beauty which ought to be cherished by the people of every land; including all such appearances as have arisen from operations not designed to create embellishment. As soon as we begin to cultivate a garden or decorate a house or an enclosure with the hope of dazzling the public eye, at that moment the spell of beauty is broken, and all the enchantment vanishes. There is something exceedingly delightful in the ornaments that have arisen spontaneously in those grounds which, after they were once reduced to tillage, have been left for many years in the primitive hands of Nature. Vain are all our attempts to imitate these indescribable beauties, such as we find along the borders of an old rustic farm, on an old roadside, or in a pasture that is overgrown with spontaneous shrubbery.

This kind of scenery is common in almost all those old roads which are not used as thoroughfares, but as avenues of communication between our small country villages. Our land is full of these rustic by-ways; and the rude scenery about them is more charming to my sight than the most highly ornamented landscapes which have been dressed by the hand of art. A part of their charm arises, undoubtedly, from their association in our minds with the simplicity of life that once prevailed among our rural population. But this is not all. I believe it arises chiefly from the almost entire absence of decoration, save that which Nature has planted with her own hands. Wherever we see a profusion of embellishments introduced by art, though they consist wholly of natural objects, we no longer feel the presence of Nature's highest charm. Something very analogous to sunshine is shut out. The rural deities do not dwell there, and cannot inspire us with a fulness of satisfaction. It is difficult to explain the cause; but when I am rambling the fields or travelling over one of these old roads with that sort of quiet rapture with which we drift along in a boat down a narrow stream through the green woods in summer, the very first highly artificial object I encounter which bears evidence of being put up for exhibition dissolves the spell, and I feel, all at once, as if I had stepped out of Paradise into the land of worldlings and vanity.

The beauty of our old roads does not consist in their crookedness, though it cannot be denied that this quality destroys their monotony and adds variety to our prospect by constantly changing our position. Neither does their beauty consist in their narrowness, though it will be admitted that this condition renders them more interesting by bringing their bushy sidewalks nearer together. Their principal charm comes from the character of their roadsides, now overgrown with all that blended variety of herbs and shrubbery which we encounter in a wild pasture. We hear a great deal of complaint of old roads, because they are crooked and narrow and because our ancestors did not plant them with trees. But trees have grown up spontaneously in many places, sometimes forming knolls and coppices of inimitable beauty; and often an irregular row of trees and shrubs of different species gives intricacy and variety to the scene.

And how much more delightful is a ride or a stroll over one of these narrow roads, than through the most highly ornamented suburbs of our cities, with their avenues of more convenient width! The very neglect to which they have been left, together with the small amount of travelling over them, has caused numberless beauties to spring up in their borders. In these places Nature seems to have regained her sovereignty. The squirrel runs freely along the walls, and the hare may be seen peeping timidly out of her burrow at their foundation, or leaping across the road. The hazel-bushes often form a natural hedgerow for whole furlongs; and the sparrow and the robin, and even some of the less familiar birds, build their nests in the green thickets of barberries, viburnums, cornels, and whortleberry-bushes that grow in irregular rows and tufts along the rough and varied embankments.

Near any old road we seldom meet an artificial object that is made disagreeable by its manifest pretensions. Little one-story cottages are frequent, with their green slope in front, and a maple or an elm that affords them shelter and shade. The old stone-wall festooned with wild grape-vines comes close up to their enclosures; and on one side of the house the garden is seen with its unpretending neatness, its few morning-glories trained up against the walls, its beds of scarlet-runners reared upon trellises formed of the bended branches of the white-birch driven into the soil, its few rose-bushes of those beautiful kinds which have long been naturalized in our gardens. When I behold these objects in their Arcadian simplicity, I lose all faith in the magnificent splendors of princely gardens. I feel persuaded that in these humble scenes dwells the highest kind of beauty, and that he is a happy man who cares for no more embellishments than his own hands have undesignedly added to the simple charms of Nature.

Let us, therefore, carefully preserve these ancient winding roads, with all their primitive eccentricities. Let no modern vandalism, misnamed public economy, deprive the traveller of their pleasant advantages, by stopping up their beautiful curves and building shorter cuts for economizing distance. Who that is journeying for pleasure is not delighted with them, as they pass on through pleasant valleys, under the brows of hills, along the banks of green rivers, or the borders of silvery lakes; now half-way up some gentle eminence that commands a view of a neighboring village, or winding round a hill, and giving us a new view of the scenes we have just passed? They are no niggardly economists of time; but they seem as if purposely contrived to present to the eye of the traveller everything that renders the country desirable to the sight; now leading us over miles bounded by old gray stone-walls, half covered with sweet-briers, viburnums, and goldenrods; then again through fragrant woods, under the brink of precipices nodding with wild shrubbery, and seeming to emulate the capricious windings of the stream in its blue course among the hills. How pleasant, when journeying, to enter a village by one of these gentle sweeps that gives us several glimpses of its scenes, in different aspects, before our arrival! How much indeed would be done for us by Nature, if we did not, in conformity with certain notions of improvement, constantly check her spontaneous efforts to cover the land with beauty!

## NOVEMBER.

A CHANGE has lately come over the face of nature; the bright garniture of field and wood has faded; the leaves have fallen to the ground, and the sun gleams brightly through the naked branches of the trees into the late dark recesses of the forest. In some years the bright hues of autumn remain unseared by frost until November has tarried with us many days. It is then melancholy to observe the change that suddenly takes place in the aspect of the woods after the first wintry night. The longer this fatal blast is deferred, the more sudden and manifest are its effects. The fields to-day may be glowing in the fairest hues of autumnal splendor. One night passes away, — a night of still, freezing cold, depositing a beautiful frostwork on our windows, — and lo ! a complete robe of monotonous brown covers the wide forest and all its colors have vanished. After this frost the leaves fall rapidly from the trees, and the first vigorous wind will nearly disrobe them of their foliage.

This change is usually more gradual. Slight frosts occur one after another during many successive nights, each adding a browner tint to the foliage and causing the different trees to shed their leaves in natural succession. Though November is the time of the general fall of the leaf, yet many trees cast off their vesture in October. But the flowering season closed with the last of the month. A few asters are still seen in the woods, and here and there on the green southern slopes a violet will look up with its mild blue eye, like a star of promise, to remind us of the beauties of the coming spring. There is a melancholy pleasure attending a ramble at this time, while taking note of the changes of the year, and of the care with which Nature provides for the preservation of her charge during the coming season of cold. All sounds that meet the ear are in harmony with our feelings. The breezes murmur with a plaintive moan, while shaking the dropping leaves from the trees, as if they felt a sympathy with the general decay, and carefully strew them over the beds of the flowers to afford them a warm covering and protection from the ungenial winter. The sear and yellow leaves eddying with the fitful breezes fill up the hollows in the pastures where slumbering lilies and violets repose, and gather around the borders of the woods, where the vernal flowers are sleeping and require their warmth and protection. There is an influence breathing from all nature in the autumn that leads us to meditate on the charms of the seasons that have flown, and prepares us by the regrets thus awakened to realize their full worth, and to experience the greater delight when we meet them once more.

There are rural sounds as well as rural sights which are characteristic of this as well as every other month of the year, all associated with the beauties and bounties of their respective seasons. The chirping of insects declines during October and dies away to silence before the middle of the present month; and then do the voices of the winter birds become more audible. Their harsh unmusical voices harmonize not unpleasantly with the murmuring of wintry winds and with the desolate appearance of nature. The water birds assemble in the harbors and are unusually loquacious; and occasionally on still evenings we hear the cackling flight of geese as they are proceeding aloft to the places of their hyemal abode. These different sounds, though unmusical and melancholy, awaken many pleasant recollections of the season, and always attract our attention.

But silence for the most part prevails in the fields and woods so lately vocal with cheerful notes. The birds that long since discontinued their songs have forsaken our territories, and but few are either heard or seen. The grasshoppers have hung their harps upon the brown sedges and are buried in a torpid sleep. The butterflies also have perished with the flowers, and the whole tribe of sportive insects that enlivened the prospect with their motions have gone from our sight. Few sounds are heard on still days, save the dropping of nuts, the rustling of leaves, and the careering of the fitful winds that often disturb the general calm. Beautiful sights and sounds have vanished together, and the rambler who goes out to greet the cheerful objects of nature finds himself alone, communing only with silence and solitude.

It is in these days of November that we most fully realize how much of the pleasure of a rural excursion is derived from the melodies that greet our ears during the vocal months of the year. Since the merry-making tenants of the groves have left them to inanimate sounds Nature seems divested of life and personality. While separated from all sounds of rejoicing and animation, we seem to be in the presence of friends who are silent and mourning over some bereavement. In the vocal season the merry voices of birds and insects give life to the inanimate objects around us, and Nature herself seems to be talking with us in our solitary but not lonely walk. But when these gay and social creatures are absent, the places they frequented are converted into solitude. No cheerful voices are speaking to us, no bright flowers are smiling upon us, and we feel like one who is left alone to mourn over the scenes of absent joys and departed friends.

But the silence to which I allude is chiefly that of the singing-birds, whose voices are the natural language of love and rejoicing. There are still many sounds which are characteristic of the month. Hollow winds are sighing through the half-leafless wood, and the sharp rustling of dry oak-leaves is heard aloft in the place of the warbling of birds and the soft whispering of zephyrs. The winds as they sweep over the shrubbery produce a shrill sound that chills us as the bleak foreboding of winter. The passing breezes have lost that mellowness of tone that comes from them in summer while floating over the tender herbs and the flexible grain. Every sound they make is sharper whether they are rustling among the dry cornfields or whistling among the naked branches of the trees. Since the forests have shed their leaves the voices of the winter birds are heard with more distinctness, and the echoes are repeated with a greater number of reverberations among the rocks and hills.

Our rural festivities are past, the harvest is gathered, and all hands are busy preparing for the comforts of the winter fireside. The days are short, and the sun at noonday looks down with a slanting beam and diminished fervor, or remains behind the cloud that often overshadows the horizon. Dark clouds of ominous forms and threatening look brood sometimes for whole days over the sullen atmosphere, through which the beams of the sun will occasionally peer, as if to bid us not wholly despair of his benignant presence. Every object in the rural world tells of the coming of snows and of the rapid passing of the genial days of autumn. The evergreens are the only lively objects that grace the landscape, and the flowers lie buried under the faded leaves of the trees that lift up their branches as if in supplication to the skies.

The spirit of desolation sits upon the hills; and in her baleful presence the northern blasts assemble on the plains, and the wintry frosts gather together in the once smiling valleys.

Such are the changes of the seasons, melancholy emblems of the vicissitudes of life. Transient is the period of youth, like the flowery month of May, and rapidly, like the flowers of summer, fade all the joys of early manhood. Our early hopes after they have finished their songs of promise vanish like the singing-birds, and the visions of our youth flit away like the insects that glitter for a few brief days and then perish forever. Yet as the pleasant things of one month are followed by those equally delightful in the next, so are the joys of youth that perish succeeded by the riper though less exhilarating pleasures of manhood. These in turn pass away to be replaced by the tranquil and sober comforts of age, as the autumnal harvest crowns the frailer products of summer. Joys are constantly alternating with sorrows, and the regrets we pour over our bereavements are softened and subdued by the new bounties and blessings of the present time. While we are lamenting the departure of one beautiful month, another no less delightful has already arrived; and the winters of our sorrow are always succeeded by vernal periods of happiness.

But to him who contemplates the works of Nature with a philosophic view, do these vicissitudes yield sources of pleasure, derived from watching the growth of the fields through all its gradations, from the bud to the flower and the leaf, and from the seedling to the perfect plant. The budding of trees, the gradual expansion of their leaves, and all the changes through which they pass until their final decay, present unfailing topics of curious and pleasing meditation. In every change that happens, he discovers a new train of reflections on the grandeur and harmony of Nature's works. Even the melancholy inspired by the autumn differs from despondency, and partakes of the character of pleasure. While lamenting the departure of flowers and the coming of snows, we are conscious that there would be a monotony in a perpetual summer, which would soon be followed by indifference; and amidst the beauties and blessings of nature, our hearts would be cloyed with luxury and sighing after unattainable happiness.

# TESTIMONY FOR THE BIRDS.

A FARMER'S boy in Ohio, observing a small flock of quails in his father's cornfield, resolved to watch their motions. They pursued a regular course in their foraging, beginning on one side of the field, taking about five rows and following them uniformly to the opposite end. Returning in the same manner over the next five rows, they continued this course until they had explored the greater part of the field. The lad, suspecting them of pulling up the corn, shot one of them and examined the ground. In the whole space over which they had travelled he found but one stalk of corn disturbed. This was nearly scratched out of the ground, but the kernel still adhered to it. In the craw of the quail he found one cutworm, twenty-one striped vine-bugs, and one hundred chinch-bugs, but not a single kernel of corn. As the quail is a granivorous bird in winter, this fact proves that even those birds that are able to subsist upon seeds prefer insects and grubs when they have their choice.

Mr. Roberts, a farmer who resided in Colesville, Ohio, was invited by a neighbor to assist him in killing some yellow-birds which, as he thought, were destroying his wheat. Mr. Roberts, not believing the birds guilty of any such mischief, was inclined to protect them. To satisfy his curiosity, however, he killed one of the yellow-birds, and found, upon opening its crop, that instead of wheat the bird had devoured the weevil, the greatest destroyer of wheat. He found in the bird's crop as many as two hundred weevils and but four grains of wheat;

and as each of those grains contained a weevil, he believed they were eaten for the sake of the insect within them. The jealousy of the Ohio farmers had prompted them in this case to destroy a family of birds, at the very time when they were performing an incalculable amount of benefit to agriculture.

The Southern farmers suspected the kildeer, a species of plover, of destroying young turnips. A writer in the "Southern Planter," alluding to this notion, declares the kildeer to be the true guardian of the turnip crop; and to prove his assertion he dissected a number of them. Their crops were found to contain no vegetable substance. Nothing was found in them save the little bug that is a well-known destroyer of turnips and tobacco-plants. They were little hopping beetles, and were rapidly increasing, because the kildeers, their natural enemies, had been nearly exterminated. "I seldom nowadays," he says, "hear the kildeer's voice. Let no man henceforth kill one except to convince himself and others that they eat no young turnips. The sacrifice of one, producing such conviction, may save hundreds of his brethren."

Insects of various kinds, in the year 1826, had become so generally destructive as to cause apprehensions for the safety of all products of the field. A correspondent of the "Massachusetts Yeoman" expressed his belief that this unusual number of injurious insects was caused by the scarcity of birds. His neighbors were astonished that everything in his garden should be so thrifty, while their plants were cut down and destroyed before they had acquired any important growth. "I have no concern about it," he replied; "my robins see to that. I preserve them from their enemies, and they preserve my garden from worms and insects. In one corner of my garden near my dwelling is a tree in which a couple of these friends of man have reared their families for three successive years. There has ever been a harmony between my birds and me." He protected all the birds that frequented his grounds, and they devoured the insects that infested them. Grasshoppers, he said, in the early stage of their existence are not bigger than flies. Ten or twelve birds would clear a whole field of them before they could do any injury to the grass-crop.

Small owls are useful as destroyers of the larger moths and nocturnal insects, and they are excellent mousers. Hon. Richard Peters, in "The Memoirs of the Philadelphia Society for Promoting Agriculture," says that all owls are destroyers of mice. A pine-tree near his house afforded a resort to about a dozen of these birds during the winter. From witnessing their operations, he concluded that a few of them, if harbored near, would clear the fields, barns, and out-houses of vermin. He says it is only the larger species that will attack poultry or do any other damage.

The inhabitants of a new country, like America, are not so well informed of the evils that follow the destruction of birds as those of old countries, who have learned from the experience of many generations the indispensable character of their services. Vincent Kollar says, if we would prevent the increase of the cockchafer, we must spare the birds that feed upon its larva. Among these he thinks the crow deserves the first place. The bird follows the plough to obtain the larva of this and other insects as they are turned out by the furrows. In gardens he walks among the plants, and wherever one has begun to wither he plunges his bill into the ground and draws out the grub. Crows do the same in the meadows, which are sometimes nearly covered with them. The American crow has the same habits; but he does not follow the plough, from his fear of the farmer's gun. Our people will not believe that the crow does anything but mischief. But John Randolph was so well convinced of the usefulness of crows, that he would not allow one to be shot upon his farm. To prevent their depredations he fed them liberally when his young corn was liable to be injured by them.

Mr. E. S. Samuels, while admitting the important services rendered by the crow as a destroyer of insects, larva, and vermin, thinks it counterbalances all its services by its habit of devouring young birds while in their nest, of which it destroys an immense number. His reasoning is logical, and I have no information that would lead me to doubt his facts. It seems probable, however, that the crow would find its time more profitably spent in exploring the fields for grubs and worms than in hunting for birds'-nests.

Nuttall, after describing the mischief done to the corn-crop by immense assemblages of crow-blackbirds in the Southern States and the hatred borne them by the farmers on that account, remarks that on their arrival their food for a long time consists wholly of those insects which are the greatest pests to the farmer. He says they familiarly follow the plough, and take all the grubs and other noxious vermin as they appear, scratching the loose soil, that none may escape. He affirms that up to the time of the harvest he has found invariably, upon dissection, that their food consists of larva, caterpillars, moths, and beetles, in such immense quantities that if they had lived they would have destroyed the whole crop.

## THE WINTER BIRDS.

WE are prone to set an extraordinary value upon all those pleasures that arrive in a season when they are few and unexpected. Hence the peculiar charms of the early flowers of spring, and of those, equally delightful, that come up to cheer the short and melancholy days of November. The winter birds, though they do not sing, are interesting on account of the season. The Chickadees and Speckled Woodpeckers, that tarry with us in midwinter and make the still, cold days lively and cheerful by their merry voices, are in animated nature what flowers would be if we saw them wreathing their forms about the leafless trees. Nature does not permit at any season an entire dearth of those sources of enjoyment that spring from observation of the external world. As there are evergreen mosses and ferns that supply in winter the places of the absent flowers, so there are chattering birds that linger in the wintry woods; and Nature has multiplied the echoes at this time, that their few and feeble voices may be repeated by lively reverberations among the hills.

To those who look upon the earth with the feelings of a poet or a painter, I need not speak of the value of the winter birds as enliveners of the landscape. Any circumstance connected with natural scenery that exercises our feelings of benevolence adds to the picturesque charms of a prospect. No man can see a little bird or quadruped at this time without feeling a lively interest in its welfare. The sight of a flock of Snow-Buntings,

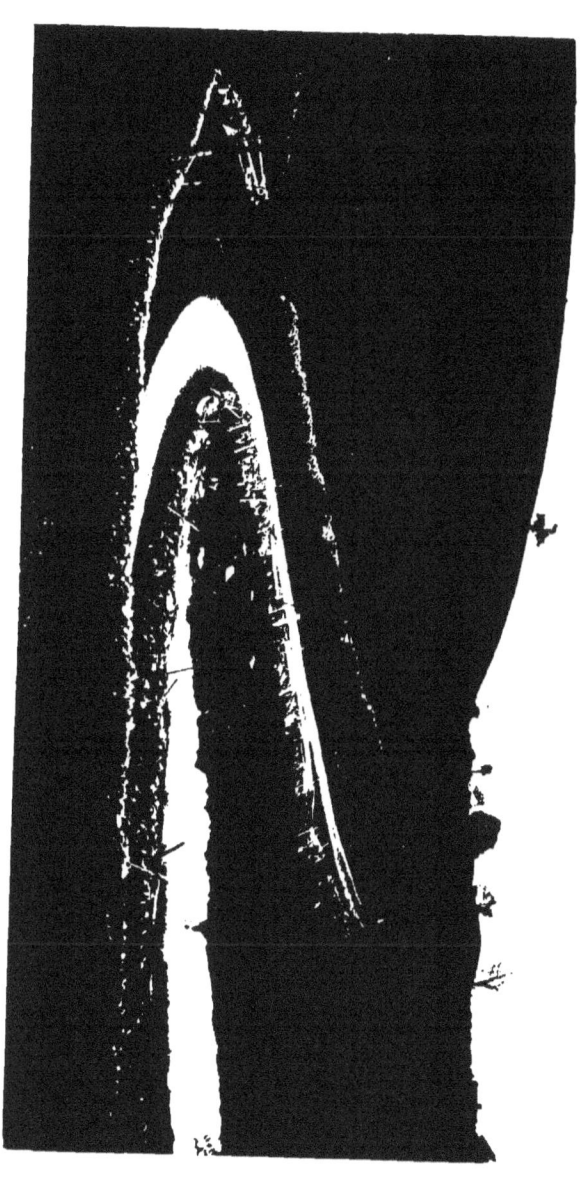

descending like a shower of meteors upon a field of grass and eagerly devouring the seeds contained in the drooping panicles that extend above the snow-drifts; of a company of Crows, rejoicing with noisy sociability over some newly discovered feast in the pine-wood; of the parti-colored Woodpeckers winding round the trees and hammering upon their trunks,—all these and many other sights and sounds are associated with our ideas of the happiness of these creatures; and while our benevolent feelings are thus agreeably exercised, the objects that cause our emotions add a positive charm to the dreary aspects of winter. These reflections have led me to regard the birds and other interesting animals as having a value to mankind not to be estimated in dollars and cents, and which is entirely independent of any services they may render the farmer or the orchardist by preventing the over-multiplication of noxious insects.

The greater number of small birds that remain in northern latitudes during winter, except the Woodpeckers, are such as live chiefly upon seeds. Those insectivorous species that gather their food chiefly from the ground, like the Thrushes and the Blackbirds, are obliged to migrate or starve. Thus the common Robins are almost exclusively insect-feeders, using fruit, that serves them rather as dessert than substantial fare. A bird that never devours seeds or grain or any farinaceous food, depending on insects and grubs that may be gathered from the surface of the ground, cannot subsist in our latitude save in mild and open winters. During such favorable seasons Robins in small parties are often seen collecting their fare of dormant insects from the open ground. The Robin, a bird that should hardly be called migratory, never proceeds any farther south than is necessary to keep him from starvation. Robins perform their migrations only as they are driven by the snow. If on any years, as sometimes happens, a large quantity of snow should cover the territory of the Middle States as early as the first of November, while north of them the ground remained uncovered, the Robins would be retarded in their journey, which is not a continued migration, and tarry with us in unusual numbers. A great many of them would perish with hunger or be reduced to the necessity of feeding on the berries of the juniper and viburnum, if they should be overtaken by snow covering a wide surface that cuts off their supplies of dormant insects.

The Quail is not so liable to be starved, because, like the common domestic hen, it is omnivorous. Quails may be kept through the winter if fed exclusively on grain. Hence, if it were not for the persecution they suffer from mankind, they would be common residents with us in winter, keeping themselves under the protection of sheds and border shrubbery and gleaning their subsistence from cornfields, and often associating with the poultry in the farm-yard. If they had been encouraged by man in a state of half-domestication, either for the use of their flesh or as consumers of grubs and insects, they might still have been common. Instead of being buried in snow in the woods, they would have crept into our barns and found safety in the hospitality of man, and would have rewarded his kindness by their invaluable services upon the farm. But man is only a half-reasoning animal. The blood of the ape still courses in his veins, rendering him incapable of understanding the value of thousands of creatures which he destroys.

The Woodpeckers and their allied families, though insectivorous, are not often distressed by the winter. Gathering all their food, consisting of larva and dormant insects, from the bark and wood of trees, the snow cannot conceal it for any perilous length of time, and only a coating of ice, that seldom outlasts more than a day or two, and covers only one side of a branch, can cause them much trouble. The quantity of their insect food is less than in summer, but the birds that winter here have about as much of it, because other species are diminished that divide with them this spoil in summer. Hence, Woodpeckers, Tomtits, and Creepers are not obliged to migrate. They simply scatter more widely over the country, instead of remaining in the woods, and thus accommodate themselves to the more limited supply of food in any given space. The Swallows and Flycatchers, that take their food in the air, are the first to migrate, because the aerial insects are vastly and suddenly diminished by the early frosts of autumn.

It is not often that we are led to reflect upon the extreme loneliness that would prevail in solitary places in winter, were all the birds to migrate at this season to a warmer climate, or to sink into a torpid state like frogs and dormice and tardy families of Swallows. But Nature, to preserve the cheerfulness of this season, has endowed certain birds with power to endure the severest cold and with the faculty of providing for their wants at a time when it would seem that there was not food enough in the hidden stores of the season to preserve them from starvation. The woodman, however insensible he may be to the charms of all such objects, is gladdened and encouraged in his toils by the sight of these lively creatures, some of which, like the Jay and the Woodpecker, are adorned with the most beautiful plumage, and are all pleasantly garrulous, filling the otherwise silent woods with frequent and vociferous merriment.

In my early days, for the supposed benefit of my health, I passed a winter in Tennessee, and, being unoccupied, except with my studies, I made almost daily journeys into the woods a few miles from the city of Nashville. It was at this season that I experienced the full power of the winter birds to give life and beauty to the faded scenes of nature. Though not one was heard to sing, they seemed as active and as full of merriment as in the early summer. The most attractive birds on this occasion were the Woodpeckers, of which several species were very numerous. Conspicuous among them was the Carpenter Bird, or Pileated Woodpecker, — a bird with rusty black plumage, a red crest and mustaches, and a white stripe on each side of the neck; one of the largest of the family. His loud croaking might be heard at any time in the deep woods, and his great size and his frequent hammering upon the resounding boles of decayed trees were very attractive.

A more beautiful but smaller species was the Redheaded Woodpecker, with head and neck and throat of crimson, and other parts of his plumage variously marked with white and changeable blue. This species, though seldom seen in Eastern Massachusetts, is a common resident in this latitude west of the Green Mountain range. The birds of this species were very numerous, and during my rambles the woods were constantly flashing with their bright colors as they flitted among the trees. They were sometimes joined by the Redbreasted Woodpecker, hardly less beautiful.

It is impossible to describe the charm which these birds gave to the otherwise solitary woods. I would sometimes remain a whole day there, watching the manners of these and hundreds of other beautiful birds, that were strangers to me, taking my dinner with the squirrels, upon the fruit of the black walnut-tree, which was strewn over all the ground. The loud croaking of the Log-cock, the cackling screams of the Red-headed Woodpecker, and the solemn tolling of the Redbreast, blended with the occasional cooing of Turtle-Doves, made a sylvan entertainment that rendered my winter rambles at this period and in these woods as interesting as any I ever pursued in summer or autumn.

In our latitude, after the first flight of snow has covered the ground, many winter birds, pressed by hunger, are compelled to make extensive forages in quest of food. Our attention is especially drawn toward them at such times, and many parties of them will visit our neighborhood in the course of the day, while, if no snow had fallen, they would have confined themselves to a more limited range. One of the most attractive sights on such occasions is caused by flocks of Snow-Buntings usually assembled in great numbers. They are chiefly seen when the snow compels them to fly from place to place in quest of food. They are not birds of ill-omen, as might be inferred from the name of *bad-weather birds* given them in Sweden; for they do not appear until the storm is over.

Few sights are more picturesque than these flocks of Snow-Buntings, whirling with the subsiding winds and moving as if they were guided by an eddying breeze, now half concealed by the direction in which they meet the rays of the sun, then suddenly flashing as with a simultaneous turn they present the under white side of their wings to the light of day. The power of these diminutive creatures to endure the cold of winter and to contend with the storm attaches to their appearance a character allied to sublimity. I cannot look upon them, therefore, in any other view than as important parts in that ever-changing picture of light, motion, and beauty with which Nature benevolently consoles us for those evils assigned by fate to all the inhabitants of the earth.

The common Snowbirds, of a bluish-slate color, are not so often seen in large compact flocks. They go usually in scattered parties, and are seen in the southern parts of New England only in winter and early spring, arriving from the northern regions late in the autumn. Wilson considers them more numerous than any other species on the American continent, swarming in multitudes over all the country down to the borders of the Gulf of Mexico. It is a marvel to him, therefore, that no part of these immense hosts should remain in the summer to breed at a latitude below that of 45° or 50°, except in the high mountain-ranges. They have many of the habits of the common Hairbird, which is by some of our countrymen supposed to be the same species, changed in appearance by the winter. Like the Hairbird, they assemble round our houses and barns, picking up seeds or crumbs of bread and other fragments of food.

They differ entirely from the Buntings, which, for distinction, are called White Snowbirds. They are quite equal to them in their power of enduring the cold and in sustaining the force of a tempest. During a snowstorm they may often be seen sporting as it were in the very whirlpool of driving snows, and alighting upon the tall weeds and sedges, and eagerly gathering their products. The Hempbird will sometimes join their parties, and his cheerful and well-known twitter may be heard, as he hurriedly flits from one bush to another, hunting for the seeds of goldenrods and asters. The cause of the migration of these birds from their native north is not probably the severe cold of those regions, but the deep snows that bury up their cereal stores at an early season. They live upon seeds; hence their forages are made chiefly in tilled lands, where weeds afford them an abundant harvest. The negligence of the tiller of the soil is therefore a great gain to the small birds, by leaving a supply of seeds in the annual grasses that grow thriftily with his crops.

Early in the spring the little Blue Snowbirds again appear, but are not so familiar as in the beginning of winter. They are often seen in a thicket in companies, warbling softly and melodiously. Nuttall says their song resembles that of the European Robin Redbreast. He also remarks that the males have severe contests when they are choosing their mates.

### THE CHICKADEE.

There are but few persons who have spent their winters in the country, who would not agree with me that to the lively notes of the Chickadee we are indebted for a great part of the cheerfulness that attends a winter's walk. His notes are not a song; but there is a liveliness in their sound, uttered most frequently on a pleasant winter's day, causing' them to be associated with all agreeable changes of the weather. The Chickadees are not seen, like Snowbirds, most numerously after a fall of snow. Their habits are nearly the same in all weathers, except that they are more prone to be noisy and loquacious on pleasant, sunny days.

The sounds from which the Chickadee has derived its name appear to be its call-notes, like the gobbling of a turkey, and enable the birds while scattered singly over the forest to signalize their presence to others of their own species. It may be observed that when the call-note is rapidly repeated, a multitude of them will immediately assemble near the place where the alarm was given. When no alarm is intended to be given, the bird utters these notes but seldom, and chiefly as it passes from one tree to another. It is probably accustomed to hearing a response, and if one is not soon heard it will repeat the call until it is answered. For as these birds do not forage the woods in flocks, this continued hailing is carried on between them, to satisfy their desire not to remain entirely alone. A similar conversation passes between a flock of chickens when scattered over a field and out of sight of one another. One, on finding itself alone, will leave its quest for food and chirp until it hears a response, when it resumes its feeding. The call-notes of this species are very lively, with a mixture of querulousness in their tone which is not unpleasant.

The Chickadee is the smallest of our winter birds. He is a permanent resident, and everybody knows him. He is a lively chatterer and an agreeable companion; and as he never tarries long in one place, he does not tire us with his garrulity. He is our attendant on all our pleasant winter walks, in the orchard and the wood, in the garden and by the roadside. We have seen him on still winter days flitting from tree to tree, with the liveliest motions and the most engaging attitudes, examining every twig and branch, and after a few sprightly notes hopping to another tree, to pass through the same manœuvres. Even those who are confined to the house are not excluded from a sight of these birds. We cannot open a window on a bright winter's morning without a greeting from one of them on the nearest tree.

Beside the note from which the Chickadee derives his name, he utters occasionally two very plaintive notes, which are separated by a true musical interval, making a third on the descending scale.

They slightly resemble those of the Pewee, and are often mistaken for them; but they are not drawling or melancholy, and do not slide from one note to another without an interval. I do not know the circumstances that prompt the bird to repeat this plaintive strain; but it is uttered both in summer and winter. It has, therefore, no connection with love or the care of the offspring. Indeed, there is such a variety in the notes uttered at different times by this bird, that if they were repeated in uninterrupted succession, they would form one of the most agreeable of woodland melodies.

The Chickadee is not a singing-bird. He utters his usual notes at all times of the year; but in the early part of summer he is addicted to a low and pleasant kind of warbling, considerably varied and wanting only more loudness and precision to entitle him to rank among the singing-birds. This warbling seems to be a sort of soliloquizing for his own amusement. If it were uttered by the young birds only, we might suppose them to be taking lessons in music and that this was one of their first attempts. I have heard a Golden Robin occasionally warbling in a similar manner.

### THE DOWNY WOODPECKER.

In company with the Chickadees, we often see two speckled Woodpeckers, differing apparently only in size, each having a small red crest. The smaller of the two is the Downy Woodpecker. The birds of this species are called Sap-suckers, from their habit of making perforations in the sound branches of trees, through the bark, without penetrating the wood, as if for the purpose of obtaining the sap. These perforations are often in two or three parallel circles around the branch, very close together, and it is probable that they follow the path of a grub that is concealed under the bark. Wilson examined many trees that were perforated in this manner, and saw evidence that they suffered no harm from it. But why the bird should be so precise and formal in his markings of the tree is a mystery not yet satisfactorily explained.

The Woodpecker, however, takes a great part of his

food from the inside of the wood or bark of the tree. Hence he is not so diligent in his examination of the outside of the branch as the Chickadee. He examines those parts only where he hears the scratching or gnawing of the grub that is concealed beneath the surface, bores the wood to obtain it, and then flies off. The Chickadee looks for insects on or near the surface, and does not confine his search to trees. He examines fences, the under parts of the eaves of houses, and the wood-pile, and destroys in the course of his foraging many an embryo moth or butterfly which would become the parent of noxious larva. The Woodpecker is often represented as the emblem of industry; but the Chickadee is more truly emblematical of this virtue, and the Woodpecker of perseverance, as he never tires when drilling into the wood of a tree in quest of his prey.

### THE HAIRY WOODPECKER.

The Hairy Woodpecker is larger than the preceding species, and their difference in size is almost the only notable distinction. It derives its name from the resemblance of some of the feathers on its back to hairs. This Woodpecker is not so often seen in summer as the smaller species, but both are about equally noticeable in winter. The nest is made in holes excavated by its own labor for this purpose in the trunk or branches of old trees. The bird commonly selects a dried and partially decayed limb, because it is more easily excavated after the hole is drilled through the outside.

### THE RED-HEADED WOODPECKER.

This is the most beautiful of our Woodpeckers and nearly as large as a Robin. It is not often seen in the

New England States, but often enough to be an acquaintance of the generality of observers. This bird, like the Robin, has gained the enmity of that conscientious class of people who cut down their fruit-trees that the boys may not have the temptation set before them to break the Eighth Commandment. The Red-headed Woodpecker seems to be in this respect more mischievous even than the Robin, for he not only takes cherries, but carries off the finest apples and feeds upon the Indian-corn when in the milk. The question is often raised, therefore, with regard to the usefulness of this bird; and it will be answered according as the person interrogated takes a view of its general utility in the economy of nature, or of its mischievousness as a consumer of fruit. Mr. George W. Rice, of West Newton, preserves his cherry-trees for the exclusive benefit of the Robins. If they do not take all the cherries, he has what they leave; but he considers the fruit more valuable for the benefit of the Robins than for any other purpose. Perhaps, however, since all men are not so wise, we should say, "Cut down all your fruit-trees and imitate the generosity of those men whom we occasionally hear of, who choose to perform this sacrifice of their own property rather than to shoot the boys."

### THE BROWN CREEPER.

Another of the companions of the Chickadee is the Brown Creeper, a bird of similar habits, often seen moving in a spiral direction around the trunks and branches of trees, and when conscious of being observed, keeping on the farther side of a branch. He is more frequently seen in winter than in summer, when he is concealed by the foliage. The different birds I have named as companions of the Chickadee often assemble by seeming accident in considerable numbers upon one tree, and, meeting perhaps more company than is agreeable to them, they make the wood resound with their noisy disputes. They may have been assembled by some note of alarm, and on finding no particular cause for it, they raise a shout that reminds us of the extraordinary vociferation with which young men and boys in the country conclude a false alarm of fire in the early part of the night. These birds are not gregarious, and, though fond of the presence of a few of their own kind, are vexed when they find themselves in a crowd.

### THE NUTHATCH.

The Nuthatch is often found in these assemblages, and may be recognized by his piercing, trumpet-like note. This bird resembles the Woodpecker in the shape of the bill, but has only one hinder toe instead of two. He is a permanent inhabitant of the cold parts of the American continent, resembling the Titmouse in his diligence and activity, and in his manners while in quest of his insect food.

There are times when even the birds I have described in this essay, that collect their food from the bark and wood of trees, are driven to great extremities. When the trees are incased in ice, which, though not impenetrable by their strong bills, prevents their discovery of their food, they are in danger of starving. At such times the gardens and barnyards are frequented by large numbers of Woodpeckers, Creepers, and Nuthatches. Driven by this necessity from their usual haunts, a piece of suet fastened firmly to the branch of a tree, at any time of the winter, would soon be discovered by them and afford them a grateful repast. I have frequently assembled them under my windows by this allurement.

## THE BLUE-JAY.

If we visit any part of the forest or live near it in the winter, we are sure to be greeted by the voice of the lively Blue Jay, another of our well-known winter birds. He has a beautiful outward appearance, under which he conceals an unamiable temper and a propensity to mischief. There is no bird in our forest that is arrayed in equal splendor. His neck of fine purple, his pale, azure crest and head with silky plumes, his black crescent-shaped collar, his wings and tail-feathers of bright blue with stripes of white and black, and his elegant form and vivacious manners render him attractive to every visitor of the woods.

But with all his beauty, he has, like the Peacock, a harsh voice. He is a thief and a disturber of the peace. He is a sort of Ishmael among the feathered tribes, who are startled at the sound of his voice and fear him as a bandit. The farmer, who is well acquainted with his habits, is no friend to him; for he takes not only what is required for his immediate wants, but hoards a variety of articles in large quantities for future use. It would seem as if he were aware when engaged in an honest and when in a dishonest expedition. While searching for food in the field or the open plain, he is extremely noisy; but when he ventures into a barn to take what does not belong to him, he is silent and stealthy and exhibits all the peculiar manners of a thief.

It would be no mean task to enumerate all the acts of mischief perpetrated by this bird, and I cannot but look upon him as one of the most guilty of the winged inhabitants of the wood. He plunders the cornfield both at seed-time and harvest; he steals every edible substance he can find and conceals it in his hiding-places; he destroys the eggs of smaller birds and devours their young.

He quarrels with all other species, and his life is a constant round of contentions. He is restless, irascible, and pugnacious, and he always appears like one who is out on some expedition. Yet, though a pest to other birds, he is a watchful parent and a faithful guardian of his offspring. It is dangerous to venture near the nest of a pair of Jays, who immediately attack the adventurer, aiming their blows at his face and eyes with savage determination.

Like the Magpie, the Jay has considerable talent for mimicry, and when tamed has been taught to articulate words like a parrot. But this talent he never exercises in a wild state. At certain times I have heard this bird utter a few notes, like the tinkling of a bell, and which, if syllabled, might form such a word as *dilly-lily;* but it is not a musical strain. Indeed, there is no music in his nature; he is fit only for "stratagems and spoils."

The Blue-Jay is a true American. He is known throughout the continent, and never visits any other country. At no season is he absent from our woods, though his voice always reminds me of winter. He is also an industrious consumer of the larger insects and grubs, atoning in this way for some of his evil deeds. I cannot say, therefore, that I would consent to his banishment, for he is one of the most cheering tenants of the groves at a season when they have but few inhabitants; and I never listen to his voice without a crowd of charming reminiscences of pleasant winter excursions and adventures at an early period of life. The very harshness of his voice has caused it to be impressed more forcibly upon my memory in connection with these scenes.

## THE CROW.

The common Crow is the representative in America of the European Rook, resembling it in many of its habits. In Europe, where land is more valuable than in this country, and where agriculture is carried on with an amount of skill that would astonish an American, the people are not so jealous of the birds. In Great Britain, rookeries are permanent establishments; and the Rooks, notwithstanding the mischief they do, are protected on account of their services to agriculture. The farmers of Europe, having learned by experience that without the aid of mischievous birds their crops would be sacrificed to the more destructive insect race, forgive them their trespasses as we forgive the trespasses of cats and dogs, who in the aggregate are vastly more destructive than birds. The respect shown to birds by any people seems to bear a certain ratio to the antiquity of the nation. Hence the sacredness with which they are regarded in Japan, where the population is so dense that the inhabitants would not consent to divide the products of their fields with the feathered race unless their usefulness had been demonstrated.

The Crow is one of the most unfortunate of birds in all his relations to man; for by the public he is regarded with hatred, and every man's hand is against him. He is protected neither by custom nor by superstition; the sentimentalist cares nothing for him as a subject of romance, and the utilitarian is blind to his services as a scavenger. The farmer considers him as the very ringleader of mischief, and uses all the means he can invent for his destruction; the friend of the singing-birds bears him a grudge as the destroyer of their eggs and their young; and even the moralist is disposed to condemn him for his cunning and dissimulation.

Hence he is everywhere hunted and persecuted, and the expedients used for his destruction are numerous and revolting to the sensibility. He is outlawed by legislative bodies; he is hunted with the gun; he is caught in crow-nets; he is hoodwinked with bits of paper smeared with birdlime, to which he is attracted by means of a bait; he is poisoned with grain steeped in hellebore and strychnine; the reeds in which he roosts are treacherously set on fire; he is pinioned by his wings, and placed on his back, and made to grapple his companions who come to his rescue. Like an infidel, he is not allowed the benefit of truth to save his reputation; and children, after receiving lessons of humanity, are taught to regard the Crow as an unworthy subject, if they were to carry the precepts taught them into practice. Every government has set a price upon his head, and public sentiment holds him up to execration.

As an apology for these atrocities his persecutors enumerate a long catalogue of misdemeanors of which he is guilty. He pillages the cornfields and pulls up the young shoots of maize to obtain the kernels attached to their roots. He destroys the eggs and young of harmless birds which are our favorites; he purloins fruit from the garden and orchard, and carries off young ducks and chickens from the farm-yard. Beside his mischievous propensities and his habits of thieving, he is accused of deceit and of a depraved disposition. He who would plead for the Crow will not deny the general truth of these accusations, but, on the other hand, would enumerate certain special benefits which he confers upon man.

In the list of services performed by this bird we find many details that should lead us to pause before we consent to his destruction. He consumes vast quantities of grubs, worms, and noxious vermin; he is a valuable scavenger, clearing the land of offensive masses of decaying animal substances; he hunts the grass-fields and pulls out and devours various cutworms and caterpillars, wherever he perceives the marks of their presence as evinced by the wilted stalks; he destroys mice, young rats, and lizards and the smaller serpents; and, lastly, he is a volunteer sentinel upon the farm and drives the Hawk from its enclosures. It is chiefly during seed-time and harvest that the depredations of the Crow are committed. During the remainder of the year his acts are all benefits, and so highly are his services appreciated by those who have written on birds that there is hardly an ornithologist that does not plead for his protection.

Let us turn our attention for a moment to his moral qualities. In vain is he condemned for cunning when without this quality he could not live. His wariness is a virtue, and, surrounded as he is by all kinds of perils, it is his principal means of self-preservation. He has no system of faith and no creed to which he is under obligations to offer himself as a martyr. His cunning is his armor; and I am persuaded that the persecutions which he has always suffered have caused the development of an amount of intelligence that elevates him many degrees above the majority of the feathered race. Hence there are few birds that equal the Crow in sagacity. He observes many things that could be understood, it would seem, only by human intelligence. He judges with accuracy from the deportment of the person approaching him, if he is prepared to do him an injury, and seems to pay no regard to one who is strolling the fields in search of flowers or for recreation. On such occasions you may come so near him as to observe his manners and even to note the varying shades of his plumage. But in vain does the gunner endeavor to approach him. So sure is he to fly at the right moment for his safety that one might suppose he could measure the distance of gunshot.

The cawing of the Crow seems to me unlike any other sound in nature. It is not melodious, though less harsh than that of the Jay. It is said that when domesticated he is capable of imitating human speech, though he cannot sing. But Æsop mistook the character of this bird when he represented him as the dupe of the fox, who gained the bit of cheese he carried in his mouth by persuading him to exhibit his musical powers. The Crow could not be fooled by any such appeals to his vanity.

The Crow is justly regarded as a homely bird; yet he is not without beauty. His coat of glossy black with violet reflections, his dark eyes and sagacious expression of countenance, his stately and graceful gait, and his steady and equable flight, all give him a proud and dignified appearance. The Crow and the Raven have always been celebrated for their gravity, — a character that seems to be caused by their black, sacerdotal vesture and by certain peculiar manifestations of intelligence in their ways and general deportment. Indeed, any one who should watch the motions of the Crow for five minutes, when he is stalking alone in the field or when he is careering with his fellows around some tall tree in the forest, would not fail to see that he deserves to be called a grave bird.

Setting aside all considerations of the services rendered by the Crow to agriculture, I esteem him for certain qualities which are agreeably associated with the charms of nature. It is not the singing-birds alone that contribute by their voices to gladden the husbandman and cheer the solitary traveller. The crowing of the cock at the break of day is as joyful a sound, though unmusical, as the voice of the Robin, who chants his lay at the same early hour. To me the cawing of the Crow is cheering and delightful, and it is heard long before the generality of birds have left their perch. If not one of the melodies of morn, it is one of the most notable sounds that herald its approach. And how intimately is the voice of this bird associated with the sunshine of calm wintry days, with our woodland excursions at this inclement season, with the stroke of the woodman's axe, with open doors in bright and pleasant weather, when the eaves are dripping with the melting snow, and with all those cheerful sounds during that period when every object is valuable that relieves the silence or softens the dreary aspect of nature!

# DECEMBER.

It is one of the most cheerful recreations for a leisure hour, to go out into the fields, under a mild, open sky, to study the various appearances of Nature that accompany the changes of the seasons, and to note those phenomena which are peculiar to a climate of frost and snow. The inhabitant of the tropics with his perpetual summer, who sees no periodical changes except the alternations of rain and drought, is deprived of a happy advantage possessed by the inhabitant of the north; and, with all the blessings of his voluptuous climate, is visited by a smaller portion of the moral enjoyments of life. In the minds of those who dwell in a northern latitude there are sentiments which are probably never felt by the indolent dweller in the land of the date and the palm; and however poetical to us may seem the imagery drawn from the pictures we have read of those blissful regions, ours is most truly the region of poetry, and of all those sentiments which poetry aims to express.

It will not be denied that in winter Nature has comparatively but few attractions; that the woods and fields offer but few temptations to ramble; and that these are such as appeal to the imagination rather than to the senses, by furnishing matter for studious reflection and calling up pleasing and poetic images. The man of phlegmatic mind sees, in all these phenomena, nothing but dreariness and desolation; while to the studious or the imaginative, every form of vegetation on the surface of the earth becomes an instructive lesson or awakens a train of imagery that inspires him, on a winter's walk, with a buoyancy not often felt in the balmy days of June. Then does he trace, with unalloyed delight, every green leaf that seems budding out for spring; and in the general stillness, every sound from abroad has a gladness in its tone not surpassed by the melodies of a summer morning.

On these pleasant days of winter, which are of frequent occurrence in our variable climate, I often indulge myself in a solitary ramble, taking note of those forms of vegetation that remain unchanged, and of the still greater number that lie folded in hyemal sleep. For such excursions the only proper time is when the earth is free from snow, which, though a beautifier of the prospect, conceals all minute objects that are strewed upon the ground or that are still feebly vegetating under the protection of the woods. The most prominent appearances are the remains of autumnal vegetation. The stalks of the faded asters are still erect, with their downy heads shaking in the breeze, which has already scattered their seeds upon the ground; and the more conspicuous tufts of the goldenrods are seen in nodding and irregular rows under the fences, or bending over the ice that covers the meadows where they grew. All these are but the faded garlands of Nature, that pleasantly remind us of the past festivities of summer, of cheerful toil, or studious recreation.

Nature never entirely conceals the beauties of the field and wood save when, for their protection, she covers them with snow. The faded remnants of last summer's vegetation may have but little positive beauty; but to the mind of the naturalist they are attractive on account of the lessons they afford and the sentiments they awaken. But there are objects in the wood which are neither faded nor leafless; and many that are leafless still retain their beauty and the appearance of life. Beside the evergreens, many of the herbs that bear the early spring flowers still retain their freshness and spread out their green leaves in the protected nook or in the recesses of the fern-covered rocks. The leaves of the wild strawberry and the cinquefoil are always green in the meadows, and those of the violet on the sheltered slope of the hill. The crowfoot and the geranium are in many places as fresh as in May; and the aquatic ranunculus and the wild cresses are brightly glowing with their emerald foliage, in the depths of the crystal watercourses that remain unfrozen beneath the wooded precipice, or in the mossy ravines of the forest.

These phenomena are doubly interesting as evidences of the continued life of the beautiful things they represent, and of the invisible and ever-watchful providence of Nature. Every step we take brings under our review other similar curiosities of vegetable life, which, by reason of their commonness, often escape our observation. On the sandy plain the slender hazel-bushes are loaded with thousands of purple aments, suspended from their flexile twigs, all ready to burst into bloom at the very first breath of spring. In the wet lands, where the surface is one continued sheet of ice, the crowded alders are so full of their embryo blossoms, that their branches seem to be hung with dark purple fruit; and the sweet-fern of the upland pastures, in still, mild weather, often faintly perfumes the atmosphere with the scent of its half-developed leaves and flowers.

But the face of Nature, at this time, is not an unfruitful subject for the poet or the painter. The evergreens, if not more beautiful, are more conspicuous than at any other season; and there are many bountiful streamlets that ripple through the woods and often in their depths find protection from the greatest cold. Around these streams the embroidering mosses are as green as the grasses in May. The water-cresses may be seen growing freshly at the bottom of their channels, and the ferns are beautiful among the shelving rocks, through which the waters make their gurgling tour. When the sun, at noonday, penetrates into these green and sheltered recesses, before the snow has come upon the earth, when the pines are waving overhead, the laurels clustering with the undergrowth, and the dewberry (evergreen-blackberry) trailing at our feet, we can easily imagine ourselves surrounded by the green luxuriance of summer. Nature seems to have prepared these pleasant evergreen retreats, that they might afford to her pious votaries a shelter during their winter walks, and a prospect to gladden their eyes, when they go out to admire her works, and pay the homage of a humble heart to the great Architect of the universe.

Nor is the season without its harvest. The bayberry, or false myrtle, in dry places gleams with dense clusters of greenish-white berries, that almost conceal the branches by their profusion; the pale azure berries of the juniper are sparkling brightly in the midst of their sombre evergreen foliage; and the prinos or black-alder bushes, glowing with the brightest scarlet fruit, and resembling at a distance pyramids of flame, are irregularly distributed over the wooded swamps. While the barberries hang in wilted and blackened clusters from their bushes in the uplands, the cranberries in the peat-meadows shine out like glistening rubies, from their masses of delicate and tangled vinery. In the open places of the woods the earth is mantled with the dark glossy green leaves of the gaultheria, half concealing its drooping crimson berries; and the mitchella, of a more curious habit, each berry being formed by the united germs of two flowers, (twins upon the same stem,) adorns similar places with fairer foliage and brighter fruit.

There is a sort of perpetual spring in these protected arbors and recesses, where we may at all times behold the springing herbs and sprouting shrubbery, when they are not hidden under the snow-drift. The American hare feeds upon the foliage of these tender herbs, when she exposes herself at this season to the aim of the gunner. She cannot so well provide for her winter wants as the squirrel, whose food, contained in a husk or a nutshell, may be abundantly hoarded in her subterranean granaries. The hare in her garment of fur, protected from the cold, feels no dread of the climate; and man is almost the only enemy who threatens her, when she comes out timidly to browse upon the scant leaves of the white clover, or the heath-like foliage of the hypericum.

But the charm of a winter's walk is derived chiefly from the flowerless plants, — the ferns and lichens of the rocks, the mosses of the dells and meres, and the trailing wintergreens of the pastoral hills. Many species of these plants seem to revel in cold weather, as if it were congenial to their health and wants. To them has Nature intrusted the care of dressing all her barren places in verdure, and of preserving a grateful remnant of summer beauty in the dreary places of winter's abode. And it is not to be wondered that, to the fanciful minds of every nation, the woods have always seemed to be peopled with fairy spirits, by whose unseen hands the earth is garlanded with lovely wreaths of verdure at a time when not a flower is to be found upon the hills or in the meadow.

Whether we are adapted to nature, or nature to us, it is not to be denied that on the face of the earth those objects that appear to be natural are more congenial to our feelings than others strictly artificial. The lichen-covered rocks, that form so remarkable a feature of the hills surrounding our coast, are far more pleasing to every man's sight than similar rocks without this garniture. All this may be partly attributed to the different associations connected with the two, in our habitual trains of thought; the one presenting to us the evidence of antiquity, the other only the disagreeable idea of that defacement so generally attendant on the progress of pioneer settlements. Hence the lichens and mosses, upon the surface of the rocks, have an expression which has always been eagerly copied by the painter, and is associated with many romantic images, like the clambering ivy upon the walls of an ancient ruined tower.

At this season, when the greater part of the landscape is either covered with snow, or with the seared and brown herbage of winter, this vegetation of the rocks has a singular interest. In summer the rocks are bald in their appearance, while all around them is fresh and lively. In winter, on the other hand, they are covered with a pale verdure, interspersed with many brilliant colors, while the surrounding surface is a comparative blank. Some objects are intrinsically beautiful, others are beautiful by suggestion, others again by contrast. This latter principle causes many things to appear delightful to the eye at one period, which at other times would, by comparison with brighter objects, seem dull and lifeless. Hence on a winter's ramble, when there is no snow upon the ground, our attention is fixed, not only upon the lichens and evergreens, but likewise on the bright purple glow that proceeds from every plat of living shrubbery which is spread out in the wild. This appearance is beautiful by contrast with the dull sombre hues of the surrounding faded herbage, and it is likewise strongly suggestive of the life and vigor of Nature. It is the vivid hue of health, and entirely unlike the hue of the same plants if they were dead or dying. It is not necessary that we should have meditated upon this sight, in order to be affected by it. We are all unconscious physiognomists of the face of Nature; and over a wide tract of country, were the vegetation blasted in autumn by some secret pestilence that had destroyed its vitality, its whole aspect would be such as to sadden every beholder, though unaware of the fatal event. As the human face in sleep wears the glow, if not the animation of waking life, so the face of Nature, in her hyemal sleep, has a glow that harmonizes with our feelings and with our sense of universal beauty.

The wildwood is always full of instruction for those who are mindful either of its general scenes or its minuter objects; and a ramble on a pleasant winter's day produces on the mind an invigorating effect that might be used as a safeguard against mental depression. The landscape, when undisfigured by art, is never without beauty, and the woods are always redolent of sweet odors that assist in perfecting the illusions that arise from agreeable sights. While the exercise thus partaken in the open air strengthens the body and improves the health, the objects presented for our contemplation are tonic and exhilarant in their action on the mind. Whatever may be the season of the year, to the student of science as well as to the lover of beauty, something is always presented to fix his attention or awaken his admiration, and he seldom returns from a woodland ramble without increased cheerfulness and a prospect of new sources of rational happiness.

# BIRDS OF THE SEA AND THE SHORE.

IN my preceding essays I have treated of birds chiefly as they are endowed with song, or have some particularly interesting trait of character. But I must not omit those birds which may be especially regarded as picturesque objects in landscape. A large proportion of these are the birds of the sea and the shore. They are not singing-birds. Nature has not provided them with the gift of song, the music of which would be lost amidst the roaring and dashing of waves. Neither do I make them the subject of my remarks as objects of Natural History, but rather as actors in the romance of Nature. I treat of them as they affect the pleasant solitudes they frequent, and increase their impressiveness chiefly by their graceful or singular flight. To the motions of birds, no less than to their beauty of plumage and to the sounds of their voices, are we indebted for a great part of the interest we feel in our native land. The more we study them, the more shall we feel that in whatever direction we turn our observations, we may extend them to infinity. There is no limit to the study of Nature. Even a subject so apparently insignificant as the flight of birds may open the eyes to new beauties in the aspects of Nature and new sources of rational delight.

Nothing can exceed the gracefulness we observe in the flight of many birds of the sea, from the Osprey, that vaults in the upper region of the clouds, down to the little Sandpiper, that charms the youthful sportsman by its merry movements and circuitous flights. These little birds belong to the tribe of Waders, which are more graceful in their walk than any that live in trees and bushes. The great length of their legs permits them to take long and unembarrassed steps and to move with great facility, nodding all the while with the most amusing gesticulations. A flock of Sandpipers on the beach where it is left open by the receding tide, employing themselves in gathering their repast of marine insects, always in motion, nodding their heads and bending their bodies as if they moved them on a pivot, now carelessly taking their food, then suddenly raising their heads upon a slight alarm, now moving in companies a short distance, then rising in a momentary flight, is, to the eye of a young sportsman, one of the most interesting sights in animated nature.

The interest we feel in these birds is caused by their picturesque assemblages in twittering flocks and by their peculiar cries. The voices of the sea-birds have a family resemblance. We can always distinguish their cries, which are shrill and piercing. Their notes are never low and could seldom be mistaken for those of land-birds. The Sandpipers afford great sport to young gunners, who overtake and surprise them upon the flats of solitary inlets when the tide is low. They arrive in dense flocks, alighting at the edge of the tide and taking the insects as they are uncovered; and the dashing of the waves close to their ranks causes them to be constantly flitting as they break at their feet. While we watch them there seems to be an active contention between them and the rippling edges of the water.

It is in winter that the picturesque movements of land-birds are most apparent. In summer and in autumn, before the fall of the leaf, birds are partially concealed by the foliage of trees and shrubs, so that the manner of their flight cannot be so easily observed. In winter, if we start a flock of them from the ground, we may watch all the peculiarities of their movements. I have alluded to the descent of Snow-Buntings upon the landscape as singularly beautiful; but the motions of a flock of Quails, when feeding in an open space in a wood or when suddenly alarmed, are equally interesting. When a Dove or a Swallow takes flight, its progress through the air is so rapid and the motions of its wings so undiscernible as to injure the beauty of its flight. We hardly observe anything so much as its rapidity. It is quite otherwise with the Quail. The body of this bird is plump and heavy and its wings short, with a peculiar concavity of the under surface when expanded. The motions of the wings are very rapid, and, having but little sweep, the bird seems to hang in the air, and is carried along moderately by a rapid vibration of the wings, describing about half a circle. Hence we see the shape of the bird during its flight.

Birds of prey are remarkable for their steady and graceful flight. The motions of their wings are slow, but they are capable of propelling themselves through the air with great rapidity. The circumgyrations of a Hawk, when reconnoitring far aloft in the heavens, are very picturesque, and have been used at all times to give character to certain landscape scenes in painting. A single picturesque attitude is sufficient to suggest a whole series of movements to one who has frequently watched them. The Raven and Crow are slow in their flight, which is apparently difficult. Hence these birds are easily overtaken and annoyed by smaller birds, which are ever watchful for an occasion to attack them without danger. Crows are not formed, like Falcons, to take their prey on the wing, and they cannot perform those graceful and difficult evolutions that distinguish the flight of birds of prey.

Small birds of the Sparrow tribe and some others generally move in an undulating course, alternately rising and sinking. The species that move in this way seldom fly to great heights, and are incapable of making a long journey without frequent intervals of rest. They perform their migrations by short daily stages. The flight of the little Sandpipers that frequent salt marshes in numerous flocks would be an interesting study. These birds are capable of sustaining an even flight in a perfectly horizontal line, only a few inches above the sandy beach. When they alight they seldom make a curve or gyration. They descend in a straight line, though obliquely. Snow-Buntings turn about, just before they reach the ground, and come down spirally. I have seen them perform the most intricate movements, like those of people in a cotillon, executed with the rapidity of arrows, when suddenly checked in their course by the discovery of a field covered with ripened grasses.

### THE KINGFISHER.

If we leave the open field and wood, and ramble near the coast of some secluded branch of the sea we may be startled by the harsh voice of the Kingfisher, like the sound of the watchman's rattle. This bird is the celebrated Alcedo or Halcyon of the ancients, who attributed to it supernatural powers. It was supposed to construct its nest upon the waves, where it was made to float like a vessel at anchor. But as the turbulence of a storm would be likely to destroy it, Nature has gifted the sitting birds with the power of stilling the motion of the winds and waves during the period of incubation. The serene weather that accompanies the summer solstice was believed to be the enchanted effect of the benign influence of this family of birds. Hence the name of Halcyon days was applied to this period of tranquillity.

It is remarkable that fable should add to these supernatural gifts the power of song, as one of the accomplishments of the Kingfisher. This belief must have been very general among the ancients, and not confined to the Greeks and Romans. Some of the Asiatic nations still wear the skin of the Kingfisher about their persons as a protection against moral and physical evils. The feathers are used as love-charms; and it is believed, if the body of the Kingfisher be evenly fixed upon a pivot, it will turn its head to the north like the magnetic needle.

The Kingfisher is singularly grotesque in his appearance, though not without beauty of plumage. His long, straight, and quadrangular bill, his short and diminutive feet and legs, his immense head, and his plumage of dusky blue, with a bluish band on the breast, and a white collar around the neck, form a mixture of the grotesque and the beautiful which, considered in connection with his singularity of habits, may account for the superstitions that attach to his history. He sits patiently, like an angler, on a post at the head of a wharf, or on the trunk of a tree that extends over the bank, and, leaning obliquely with extended head and beak, he watches for his finny prey. There, with the light-blue sky above him and the dark-blue waves beneath, nothing on the surface of the water can escape his penetrating eyes. Quickly, with a sudden swoop, he seizes a single fish from an unsuspecting shoal, and announces his success by the peculiar sound of his rattle.

### THE SPOTTED TATTLER.

A very interesting bird inhabiting the shores of seas and lakes is the Peetweet, or Spotted Tattler. The birds of this species breed in all parts of New England, arriving soon after the first of May, and assembling in occasional twittering flocks, skimming along the edges of some creek or inlet, most numerously after the tide has left the beach. In their circuitous flights they follow all the inequalities of the coast. It is amusing to watch their ways when they are preparing for incubation, restless and anxious, and uttering their lively and plaintive cry, like the syllables *peet-weet*, repeating the last with the rising inflection. They resemble the notes of the little Wood-Sparrow, when repeated many times in succession, except that the Tattler utters them without increasing their rapidity or varying their tone. These notes approach more nearly to music than those of any other bird of the sea or the shore.

The Tattlers build in the meadows among the rushes, sometimes in a tilled field and very near human dwellings, where they are seen roaming about with their young, like a hen with her chickens, searching for worms and grubs. They are very liable to be shot, while attracting attention by their lively motions and their low and musical flight. The young follow the parent as soon as they are hatched, when their downy plumage is of an almost uniform light-grayish color. If surprised, they immediately hide themselves among the herbage, while the parent by her motions and cries endeavors to draw attention exclusively to herself.

The birds of this species have been so wantonly and mercilessly hunted by gunners of all ages, that they have become extremely shy, and have lost all confidence in man. Yet, if they were harbored and protected from annoyance and danger, they would grow tame and confiding, and our fields and gardens would be full of them. A brood of them following the hen would be indefatigable hunters of insects in pastures and tilled lands. A few pairs with their young would perform incalculable service on every farm, and if encouraged and protected, would soon reward us with their confidence and their services. These little birds are incapable of doing any mischief, even if there were fifty of them on every farm. They take no fruit; they do not bite off the tops of tender herbs, like poultry; they are interesting in their ways; and the only cause of their scarcity is the destruction of them by gunners.

### THE UPLAND PLOVER.

This is a species allied to the Peetweet, and well known by the name of Hill-Birds. They are of a solitary habit, not to be compared in utility and interest with the little Peetweet. They are seldom seen in flocks. We know them chiefly by their notes, which are familiar to all as heard at dawn or early evening twilight. These melancholy whistling notes are uttered as they pass from their feeding-places, while flying at a great height, and the hour of darkness when they are heard, and their plaintive modulation, render them the most striking sounds of a late summer evening.

### THE GULL.

Among the birds which are most conspicuous about our coast, I should mention the Gulls. They are not very interesting birds; but their screaming voices remind us of their habitats, and their picturesque motions are familiar to all who are accustomed to the sea-shore. They associate in miscellaneous flocks, containing often several species, and enliven the hour and the prospect by their manœuvres and their peculiar cries. The Gull is distinguished by its small and lean body, which is covered with a great quantity of feathers. Its wings and head are very large, all uniting to give the bird a false appearance of size. Hence, I suppose, originated the word, when used to imply deception. The sportsman who for the first time has shot one of these birds, expecting to find it large and plump, and discovers only a miserable lean carcass imbedded in a large mass of feathers, is said to be gulled.

### THE LOON.

I must not conclude without mentioning the Loon, one of the most romantic of birds, the Hermit of our northern lakes, and so exceedingly shy that it is rarely seen except at a great distance. This bird belongs to the family of Divers, so called from their habit of disappearing under the water at the moment when they catch a glimpse of any human being. The Loon inhabits the northern parts of Europe and North America, and is occasionally seen and heard in the lakes of New England, but chiefly now in those of Northern Maine. As population increases, this species retires to more solitary places.

In allusion to the scream of this bird, Nuttall says: "Far out at sea in winter and in the great northern lakes, I have often heard on a fine, calm morning the sad and wolfish call of the solitary Loon, which like a dismal echo seems slowly to invade the ear, and rising as it proceeds dies away in the air. This boding sound to the mariner, supposed to be indicative of a storm, may be heard sometimes for two or three miles, when the bird itself is invisible or reduced almost to a speck in the distance. The aborigines, almost as superstitious as sailors, dislike to hear the cry of the Loon, considering the bird, from its shy and extraordinary habits, as a sort of supernatural being. By the Norwegians it is with more appearance of reason supposed to portend rain."

# OLD HOUSES.

WHEN journeying in the country, who has not occasionally felt that the sight of the finest houses and the most highly ornamented grounds does not affect the mind with the greatest pleasure? We are soon tired of objects, however beautiful, that produce no other effect than to excite an agreeable visual sensation. Something that affords a pleasing exercise for the sympathies and the imagination must be blended with all scenes of beauty, or they soon become vapid and uninteresting. When we first enter the interior of a spacious dome, which is surrounded with colored glass windows, the physical sensation of beauty thus produced may detain us a few moments with extreme pleasure. But a frequent repetition of these visits would cause the spectacle to become tiresome, because it excites the eye without affecting the mind. The very opposite effect would be produced by visiting a gallery of paintings, because there is no end to the ideas and images which these works of genius may suggest.

In like manner, when travelling among the scenes of nature and art, many a highly ornamented house passes before our eyes without making any better impression upon the mind than that which is produced by examining the plates of fashions in the window of a tailor's shop. As we proceed farther into the country we presently encounter a scene that awakens a different class of emotions, that seem to penetrate more deeply into the soul. An old house, containing two stories in front, with the back roof extending almost to the ground, is seen half protected by the drooping branches of a venerable elm. A Virginia creeper hangs in careless festoons around the low windows, and a white rose-bush grows luxuriantly over the plain board fence that encloses the garden. The house stands a few rods back from the road, and is surrounded in front and on one side by an extensive grass-plat, neatly shorn by the grazing animals while sauntering on their return from pasture. An old barn is near; and the flocks and the poultry seem to enjoy an amount of comfort which we might look for in vain in the vicinity of a more ornate dwelling-house.

There is an appearance of comfort and freedom about this old house that renders it a pleasing object to almost every eye. No one can see it without calling to mind the old-fashioned people whom we always suppose to be its occupants. About it and around it we see no evidences of that constraint to which the in-dwellers and visitors of some more fashionable houses must be doomed. The exterior is associated with its interior arrangements, no less than with the scenes around it. We see, in the mind's eye, the wide entry into which the front door opens, the broad and angular staircase, the window in the upper entry, that looks out upon a rustic landscape dotted with fruit-trees, and patches of ploughed land alternating with green meadow. By the side of the staircase, on the lower floor, stands an ancient clock, whose loud striking and slow stroke of the pendulum are associated with the old style of low-studded rooms. Perhaps by studying the cause of the pleasant emotions with which we contemplate this old house, we may arrive at the knowledge of a principle that may be turned to advantage in regulating our own and the public taste.

The charm of these old houses, which are marked by neatness and plainness, and by an absence of all pretension, is founded on the natural yearning of every human soul after freedom and simplicity. In them we behold the evidences of a mode of life, which, if we could but rid our hearts of a little insanity, we should above all choose for ourselves. The human heart naturally attaches itself to those scenes in which it would be free to indulge its own natural fancies. But there is a habit stronger than nature, derived from our perverted education, that causes us to choose a part that will excite the envy of our neighbors, in preference to one that would best promote our own happiness. Hence a man chooses to be embarrassed with expenses above his pecuniary condition, for the vain purpose of exciting admiration, rather than to gratify his own tastes in the enjoyment of greater freedom and a more humble and frugal mode of life.

In vain does the worshipper of fashion, by planting an ornate dwelling-house in the heart of a forest, endeavor to add to it the charm of a rustic cottage in the woods. The traveller, as he beholds its proud ornaments glittering through the trees, sees nothing of that charming repose which, like a halo of beauty, surrounds the cottage of the rustic. He perceives in it the expression of a striving after something that is incompatible with its affectations. There may be a true love of nature among the inmates of this house. But they cannot consent wholly to relinquish that bondage of fashion which overpowers their love of freedom and simplicity, as the appetite of the inebriate causes him, in spite of his better resolutions, to turn back to the cup that is destroying him. Nature may harmonize with elegance, refinement, and grandeur, but not with pretence. The rural deities will not make their haunts near the abode of vanity; and the Naiad, when she sees her rustic fountain destroyed, turns sorrowfully away from the spouting foam of a *jet d'eau*.

There may be more true love of Nature in the inmates of this ambitious dwelling than in those of the rustic cottage; but the former gives no evidence of this love, if it is built in a style expressive of that folly which is continually drawing us away from Nature and happiness. Place them both in a picture, and the fashionable house excites only the idea of coxcombry, while the rustic cottage charms all hearts. Is it not possible to borrow this indescribable charm and add it to our country residences? Not until the builder or designer has become as one of these cottagers in the simplicity of his heart, and is content to forget the world when he is planning for his retirement. Then might the traveller pause to contemplate with delight a house in which the absence of all affectation renders doubly charming those rural accompaniments, in which the wealth of the owner, if he be wealthy, is detected only by the simple magnificence of his grounds, and his taste displayed by the charm which art has added to Nature, without degrading her fauns and her hamadryads into mere deities of the boudoir.

These old houses with a long back roof are not the only picturesque houses among our ancient buildings; but no other style seems to me so entirely American. Wherever we journey in New England, we find neat little cottages of one story, some with a door in front dividing the house into two equal parts, some with a door at the side of the front, and a vestibule with a door at the opposite end. It is common, when you meet with an old cottage of this style in the less frequented roads in the country, to see an elm standing in front, shading a wide extent of lawn. Sometimes there may be merely an apple-tree or pear-tree for purposes of shade. A rose-bush under one of the windows, bearing flowers of a deep crimson, and a lilac at the corner of the garden near the house, are perhaps the only shrubbery. These humble dwellings are the

principal attraction in some of our old winding roads, and they are remembered in connection with many delightful rural excursions. The rage that has possessed the sons of the original occupants of these cottages for putting up pasteboard imitations of something existing partly in romance and partly in the imagination of the designer, has destroyed the rurality of many scenes in our old country villages.

Any marks of pretension, or of striving after something beyond the supposed circumstances of the occupants of a house, are disagreeable to the spectator. Could the sons of the old-fashioned people who occupy plain dwellings have labored to preserve their simplicity and rustic expression, combined with a purer style of architecture, the effect would have been exceedingly pleasing. They have done the very opposite of this. Ambitious to exclude from their houses everything that would be remotely suggestive of the simple habits of rural life, they have endeavored to make them look as much as possible, with one hundredth part of the cost, like the villa of a nobleman. So many of these ambitious cottages have been reared on many of our old roads, as to have entirely destroyed that picturesque beauty which made almost every route a pleasant landscape. The road, once covered on all sides with those rural scenes that charm every lover of the country, has become as tame as one of those new-made roads, laid out by speculators on devastated ground, to be sold in lots under the hammer of the auctioneer.

The New England people have been repeatedly characterized as wanting in taste; and this deficiency is supposed to be exemplified in the entire absence of ornamental work about our old houses and their enclosures. It is a maxim that a person who is deficient in taste always runs to an extreme in the use of ornaments, whenever he attempts to use them. Hence the profusely decorated houses of the present generation do not evince any positive improvement in taste, when compared with those of their predecessors. They are simply a proof that the people of the present time have more ambition. That want of taste, which former generations exhibited by their entire disregard of ornament, is manifested in their successors by their profuse and indiscriminate use of it. The present house is no longer a thing to be loved. We cease to look upon it with affection. It is a glittering thing that merely pleases the eye, but awakens no delightful sentiment.

The object of these remarks is not to deride wealth, but to condemn the ostentation of wealth that does not exist, and the use of a house as a false advertisement of the personal importance of its owner. An intelligent man of wealth would reject these meretricious decorations as the mere sham substitute for something better which he would adopt because he could afford it. The false taste which is here censured is mere architectural hypocrisy. My purpose is to analyze certain of our emotions and sentiments, and to prove thereby that the man who builds a showy house gains no admiration, and essentially mars his own happiness. Why do we contemplate with the purest delight a simple cottage in a half-rude, half-cultivated field, except that it gives indications of something adapted to confer happiness upon its inmates? The rustic well, with a long pole fastened to a lever, by which the bucket is raised; the neat stone-wall or iron-gray fence that marks the boundary of the yard; the old standard apple-trees dotted about irregularly all over the grounds; the never-failing brook following its native circuitous course through the meadow, — all these objects present to the eye a scene that is strongly suggestive of domestic comfort and happiness.

Let us not, in our zeal for rearing something beautiful, overlook the effect of these venerable relics of the more simple mode of life that prevailed fifty years since. Let us not mistake mere glitter for beauty, nor the promptings of vanity for those of taste. Let us beware, lest in our passion for improvement, without a rational aim, we banish simplicity from the old farm, and allow fashion to usurp the throne of Nature in her own groves. Far distant be the time when the less familiar birds of our forest are compelled to retire beyond the confines of our villages, and when the red-thrush is heard only in a few solitary places, mourning over that barbarous art which has destroyed every green thicket of native shrubbery, where alone she makes her haunts. This rage for foreign shrubbery is fatal to the birds, each species being dependent on certain native trees and shrubs for subsistence and protection. By eradicating every native coppice, and planting exotics in their place, we may as effectually banish the thrushes, and many other species of warblers, from our territories as by constantly shooting them.

Another style of old houses is a square with a hipped roof, usually of two stories. This is a little more pretentious than the others I have described, and is more frequently seen with an ornamental fence in front, after the present fashion. Hence it is less attractive than some of the more primitive houses. A more pleasing style is a nearly square building of one story, with a curb roof, having the front door at the extreme end of the front, and a vestibule on one side, formed by extending the rear half of the house a few feet, with only half a roof, making the door in the vestibule and the front door face the same way. Modern improvers say there is no beauty in these old houses. As well might they say there is no beauty in an old tree, unless it is nicely trimmed and whitewashed. More charming to the sight is a humble two-story house, unadorned by a single artistical decoration, with a venerable old tree in front and a wide extent of lawn, than the most showy house in the modern filigree style, with its narrow enclosures, its stiff spruces, and its ornamental fence that seems purposely designed to shut out Nature.

One principal charm of a cottage consists in the rural appurtenances around it; and the less inexpressive architectural ornament there is about it the greater is this charm. It is true there is a style of building which is always pleasing to the eye, and another which is either offensive or unattractive. A good style differs from a bad style chiefly in suggesting, by its external appearance, all those exterior and interior arrangements which serve to make it a happy and comfortable residence. This is the principal beauty which is desirable in a dwelling in order to produce the most charming effect. There are certain ornaments the utility of which is not apparent; but everything added externally to a house, in accordance with the rule of proportions, that suggests to the mind an additional comfort or convenience, renders it more pleasing to the sight. Hence a plain, square house, without a single projection, is not so pleasant to look upon as another, whose wings and vestibules, under separate roofs, exhibit at once to the mind the conveniences within. A neatness and elegance of finish would improve it still further; but any inexpressive ornaments would spoil it. There is a class of ornaments, however, which are beautiful from suggesting something, independent of actual utility, that is agreeable to the imagination.

I would venture to affirm that the more showy the house, other things being equal, the less pleasure does it confer upon its owner or occupant. A perpetual glitter soon tires upon the eye and wearies the mind. There is a want of what painters call repose in a building that is excessively ornate; and the occupants of such a house must feel less tranquil satisfaction in it than in one of equal convenience, which is furnished only with such ornaments as have been denominated chaste. Chaste pleasures are those which are attended by no disgust and bring no repentance; and chaste ornaments resemble them in this respect, by giving permanent satisfaction, and by causing no fatigue to the eye or repentance to the mind. There is a stronger analogy between these two things than any one who has not reflected upon the subject can be aware of. It is safe to assert that any particular style of building and grounds, which serves in the highest degree to promote the happiness of the permanent occupants, will confer the most enduring pleasure upon the beholder.

We frequently admire without one spark of affection, and love with deep affection what we do not admire. But more pleasure springs from love than from admiration; and when people madly relinquish those humble scenes and objects which they love, to obtain those which shall glitter in the public eye, tickle their own vanity and excite the envy of their neighbors, they commit a greater error than the most bitter declaimer against pride has generally imagined. I am far from believing the paradox, maintained by Rousseau, that man is more happy in a state of nature than in a civilized state. This author, in his efforts to grasp at an important truth, reached beyond it. That great truth I believe to be this: that the more we extend and cultivate the moral and intellectual advantages and refinements of civilization, while we tie ourselves down to the simple habits of rustic life, the greater will be the sum of our happiness.

# JANUARY.

POETS in all ages have sung of the delights of seed-time and harvest, and of the voluptuous pleasures of summer; but when treating of winter, they have confined their descriptions to the sports of the season rather than to the beauties of Nature. Winter is supposed to furnish but few enjoyments to be compared with those of summer; because the majority of men, being oppressed by too many burdens, naturally yearn for a life of indolence. I will not deny that the pleasures derived from the direct influence of Nature are greatly diminished in cold weather; there are not so many interesting objects to amuse the mind, as in the season when all animated things are awake, and the earth is covered with vegetation; but there are many pleasant rural excursions and invigorating exercises which can be enjoyed only in the winter season, and for which thousands of our undegenerate yeomanry would welcome its annual visit.

The pleasures of a winter's walk are chiefly such as are derived from prospect. A landscape-painter could be but partially acquainted with the sublimity of terrestrial scenery, if he had never looked upon the earth when it was covered with snow. In summer the prospect unfolds such an infinite array of beautiful things to our sight, that the sublimity of the scene is hidden beneath a spectacle of dazzling and flowery splendor. We are then more powerfully attracted by objects of beauty that charm the senses than by those grander aspects of Nature that awaken the emotion of sublimity. In winter, the earth is divested of all those accompaniments of scenery which are not in unison with grandeur. At this period, therefore, the mind is affected with nobler thoughts; it is less bewildered by a multitude of fascinating objects, and is more free to indulge itself in a serious train of meditations.

The exhilaration of mind attending a winter walk in the fields and woods, when the earth is covered with snow, surpasses any emotion of the kind which is produced by the appearance of Nature at other seasons. We often hear in conversation of the invigorating effects of cold weather; yet those few only who are engaged in rural occupations, and who spend the greater part of the day in the open air, can fully realize the amount of physical enjoyment that springs from it. I can appreciate the languid recreations of a warm summer's day. When one is at leisure in the country he cannot fail to enjoy it, if he can take shelter under the canopy of trees or in the deeper shade of the forest. But these languid enjoyments would soon become oppressive and monotonous; and the constant participation of them must cause one gradually to degenerate into a mere animal. The human mind is constituted to feel positive pleasure only in action. Sleep and rest are mere negative conditions, to which we submit with a grateful sense of their power to fit us for the renewed exercise of the mind and the body.

In our latitude, at the present era January is usually the month of the greatest cold; and in severe weather there is a general stillness that is favorable to musing. The little streamlets are frozen and silent, and there is hardly any motion except of the winds, and of the trees that bend to their force. But the works of Nature are still carried on beneath the frost and snow. Though the flowers are buried in their hyemal sleep, thousands of unseen elements are present, all waiting to prepare their hues and fragrance, when the spring returns and wakes the flowers and calls the bees out from their hives. Nature is always active in her operations; and during winter are the embryos nursed of myriad hosts, that will soon spread beauty over the plains and give animation to the field and forest.

Since the beauties of summer and autumn have faded, Nature has bestowed on earth and man a brilliant recompense, and spread the prospect with new scenes of beauty and sublimity. The frozen branches of the trees are clattering in the wind, and the reed stands nodding above the ice and shivers in the rustling breeze. But while these things remind us of the chills of winter, the universal prospect of snow sends into the soul the light of its own perfect purity and splendor, and makes the landscape still beautiful in its desolation. Though we look in vain for a green herb, save where the ferns and mosses conceal themselves in little dingles among the rocks, yet the general face of the earth is unsurpassed in brilliancy. Morning, noon, and night exhibit glories unknown to any other season; and the moon is more lovely when she looks down from her starry throne and over field, lake, mountain, and valley, emblems the tranquillity of heaven.

It is pleasant to watch the progress and movements of a snow-storm while the flakes are thickly falling from the skies, and the drifts are rapidly accumulating along the sides of the fences and in the lanes and hollows. The peculiar motion of the winds, while eddying and whirling over the varied surface of the ground, is rendered more apparent than by any other phenomenon. Every curve and every irregular twisting of the wind is made palpable, to a degree that is never witnessed in the whirling leaves of autumn, in the sand of the desert, or in the dashing spray of the ocean. The appearance is less exciting when the snow descends through a perfectly still atmosphere, but after its cessation we may witness a spectacle of singular beauty. If there has been no wind to disturb the snow-flakes as they were deposited on the branches of the trees, to which they adhere, they hang from them like a drapery of muslin; then do we see throughout the woods the mimic splendor of June; and the plumage of snow suspended from the branches revives in fancy's eye the white clustering blossoms of the orchards in early summer.

Sometimes when the woods are fully wreathed in snow-flakes, and the earth is clothed in an interminable robe of ermine, the full moon rises upon the landscape and illumines the whole scene with a kind of unearthly splendor. If we wake out of sleep into a sudden view of this enchanted scene, though the mind be wearied and depressed, it is impossible, without rapture, to contemlate the etherial prospect. The unblemished purity of the snow-picture, before the senses are awakened to a full consciousness of our situation, glows upon the vision like a scene from that fairy world which has often gleamed upon the soul during its youthful season of romance and poetry. And when the early rays of morning penetrate these feathery branches and spread over the white and spotless hills of snow a rosy tinge, like the hues that burnish the clouds at sunset, and kindle amid the glittering fleece that is wreathed around the branches all the changeable colors of the rainbow, we are tempted to exclaim that the summer landscape with all its verdure and fruits and flowers was never more lovely than this transitory scene of beauty. Yet the brilliancy of this spectacle, like the rainbow in heaven, passes away almost while we are gazing on its fantastic splendor. A brisk current of wind scatters from the branches, like the fading leaves of autumn, all the false honors that have garlanded the forests, and in an hour they have disappeared forever.

Beside the pleasing objects already described as peculiar to the season, there are many beautiful appearances formed by the freezing of waters and the crystallization of vapors which one can never cease to examine with delight. One of the most brilliant spectacles of this kind is displayed on a frosty morning, after the prevalence of a damp sea-breeze. The crystals, almost imperceptibly minute, are distributed, like the delicate filaments of the microscopic mosses, over the withered herbs and leafless shrubbery, creating a sort of mimic vegetation in the late abodes of the flowers. Vast sheets of thin ice overspread the plains, beneath which the water has sunk into the earth, leaving the vacant spots of a pure whiteness, and forming hundreds of little fairy circles of a peculiarly fantastic appearance. The ferns and sedges that lift up their bended blades and feathers through the plates of ice, coated with millions of crystals, resemble, while sparkling in the rays of the sun, the finest jewelry. After a damp and frosty night, these appearances are singularly beautiful, and all the branches of the trees glitter with them as if surrounded with a network of diamonds.

These exhibitions of frostwork are still more magnificent at waterfalls, where a constant vapor arises with the spray and deposits upon the icicles that hang from the projecting rocks a plumage, resembling the finest ermine. Some of the icicles, by a constant accumulation of water which is always dripping from the crags, have attained the size of pillars, that seem almost to support the shelving rocks from which they are suspended. The foam of the water has been frozen into large white masses, like a snow-bank in appearance, but as solid as ice. The shrubs that project from the crevices of the rocks are clad in a full armor of variegated icicles; and when the slanting rays of the sun penetrate into these recesses, they illuminate them with a dazzling brilliancy; and it seems as if the nymphs, that sit by these fountains, had decorated them as the portals to that inner temple of Nature, whence are the issues of all that is lovely and beautiful on earth.

Thus, when the delightful objects of summer have perished, endless sources of amusement and delight are still provided for the mind and the senses. Though the singing-bird has fled from the orchard and the rustling of green leaves is heard no longer in the haunts of the little mountain streams, there are still many things to attract attention by their beauty or their sublimity. Whether we view the frosts that decorate the herbage in the morning, or the widespread loveliness of the snow on a moonlight evening, the sublimity of heaven seems to rest upon the face of the earth and we behold with rapt emotions every terrestrial scene. The universe, full of these harmonies, yields never-ending themes for study and meditation, to absorb and delight the mind that is ever searching after knowledge, and to raise the soul above the clods of the valley to that invisible Power that dwells throughout all space.

I never listen to the shrill voice of the woodpecker, within the deep shelters of the forest, or to the lively notes of the chickadee, which alternate with the sound of winds among the dry rustling leaves, without feeling a sudden and delightful transport. I cannot help indulging the fancy, that Nature has purposely endowed these active birds with a hardihood almost miraculous, to endure the severity of winter, that they might always remain to cheer the loneliness of these wintry solitudes. For no clime or season has Nature omitted to provide blessings for those who are willing to receive them, and in winter, wheresoever we turn, we find a thousand pleasant recompenses for our privations. The Naiad still sits by her fountain, at the foot of the valley, distributing her favors to the husbandman and his flocks; and the echoes still repeat their voices from the summits of the hills and send them over the plains, with multiplied reverberations, to cheer the hearts of all living creatures.

# FACTS THAT PROVE THE UTILITY OF BIRDS.

THE consequences which have followed the destruction of birds in many well-authenticated instances are sufficient to demonstrate their utility. Professor Jenks mentions a case communicated by one of his female correspondents. In former times, as she had been told by her father, an annual shooting-match was customary on election day in May. On one of these occasions, about the year 1820, in North Bridgewater, Mass., the birds were killed in such quantities that cart-loads of them were sent to farmers for compost. Then followed a great scarcity of birds in all that vicinity. The herbs soon showed signs of injury. Tufts of withered grass appeared and spread out widely into circles of a seared and burnt complexion. Though the cause and effect were so near each other, they were not logically put together by the inhabitants at that time. Modern entomology would have explained to them the cause of the phenomena, by the increase of the larva of insects which were previously kept in check by the birds destroyed at the shooting-match.

After the abolition of the game-laws in France, at the close of the last century, the people, having been accustomed to regard birds as the property of great land-owners, destroyed them without limit. Every species of game, including even the small singing-birds, was in danger of extermination. It was found necessary to protect them by laws that forbade hunting at certain seasons. The most serious evils were the consequence. The farmers' crops were destroyed by insects, and the orchards produced no fruit. It is only by such unfortunate experience that men can learn that the principal value of birds does not consist alone in their flesh or in their power of conferring pleasure by their songs.

Some years ago, in Virginia and North Carolina, several tracts of forest were attacked by a malady that caused the trees to perish over hundreds of acres. A traveller passing through that region inquired of a countryman if he knew the cause of the devastation. He replied that the mischief was all done by the woodpeckers, and though the inhabitants had killed great numbers of them, there still remained enough to bore into the trees and destroy them. The traveller, not satisfied with this account, made some investigations, and soon convinced them that the cause of the mischief was the larva of a species of Buprestis, which had multiplied without limits. This larva was the favorite food of the woodpeckers, which had congregated in that region lately on account of its abundance. He showed them that they were protecting the real destroyers of the forest by warring against the woodpeckers, which, if left unmolested, would soon eradicate the pest sufficiently to save the remaining timber. Birds become accustomed to certain localities, and if by any accident they should be exterminated in any one region insects of all kinds will increase, until the birds that consume them are slowly attracted to them from other parts.

In the year 1798, in the forests of Saxony and Brandenburg, the greater part of the trees, especially the conifers, died, as if struck at the roots by some secret malady. The foliage had not been attacked, and the trees perished without any manifest external cause. The Regency of Saxony sent naturalists and foresters to investigate the conditions. They proved the malady to be caused by the multiplication of a species of lepidopterous insects, which had in its larva state penetrated into the wood. Wherever a bough of fir or pine was broken, the larva was found, and had often hollowed it out even to the bark. The report of the naturalists declared that the extraordinary increase of this insect was owing to the entire disappearance of several species of titmouse, which for some years past had not been seen in that region.

According to an account given by Buffon, the Isle of Bourbon, where there were no grackles, was overrun with locusts imported in the eggs contained in the soil which with some plants had been brought from Madagascar. The Governor-General, as a means of extirpating these insects, caused several pairs of Indian grackles to be brought into the island. When the birds had considerably increased, some of the colonists, seeing them very diligent in the newly sown fields, imagined them in quest of the grain, and reported that they did more mischief than good. Accordingly they were proscribed by the Council, and in two hours after their sentence was pronounced, not a grackle was to be seen on the island. The people soon had cause for repentance. The locusts multiplied without check and became a pest. After a few years of experience, the grackles were again introduced, and their breeding and preservation were made a state affair. The birds multiplied and the locusts disappeared.

Kalm, a pupil of Linnæus, remarks in his "Travels in America," that after a great destruction of purple grackles for the legal reward of threepence per dozen, the Northern States in 1749 experienced a total loss of the grain and grass crops from the devastation of insects and their larva. The crows of North America were some years since so nearly exterminated, to obtain the premiums offered for their heads, that the increase of insects was alarming, and the States were obliged to offer bounties for the protection of crows. The same incidents have repeatedly happened in other countries, and ought to convince any reasoning mind that all the native species of insectivorous birds are needful, and that one or any number of species cannot perform the work which would have been done by the species that is extirpated.

"An aged man" of Virginia remarks, in "The Southern Planter" of 1860, that since his boyhood there has been a rapid decrease in the numbers of birds and a proportional increase of insects. Since their diminution great ravages have been committed on the farmers' crops by clover worms, wire-worms, cut-worms, and on the wheat crops particularly by chinch-bugs, Hessian flies, joint-worms, and other pests. All this is owing, he thinks, to the destruction and the scarcity of birds. He alludes particularly to the diminution of woodpeckers as a public calamity. He has known a community of red-headed woodpeckers to arrest the destructive progress of borers in a pine forest. He mentions the flicker or widgeon woodpecker — a common bird in New England — as the only bird he ever saw pulling out grubs from the roots of peach-trees. May not this habit of the flicker, which is a very shy bird because he is hunted for his flesh, be the cause why apple-trees that grow near a wood are not affected by borers?

The alarming increase of grasshoppers in some parts of the Western States, is undoubtedly the consequence of the wholesale destruction of quails, grouse, and other birds in that region.

# BIRDS OF THE FARM AND THE FARM-YARD.

It is not easy to explain why certain species of birds and other animals are susceptible of domestication, while others resist all efforts to inure them to artificial habits. The mystery is increased when we consider that individuals of a species which cannot be domesticated may, when reared in a cage, be made as tame as the tamest of our domestic birds. There are certain families of which several species have been domesticated. This is true of the Gallinaceous tribe and of the Anseres. Of the former are the Cock, the Turkey, the Pintado, the Peacock. Of the Anseres, there are two or three species of Goose and several species of Duck. Several of the Pigeon tribe may be domesticated. The Rook and the House Sparrow of England may also be regarded as in a state of at least partial domestication. The species among our birds that comes nearest the Rook in its habits is the Purple Grackle. That, as population thickens, the Grackles will assume more and more of the habits of a domestic bird, seems not improbable, especially if they should be protected for their valuable services to agriculture.

THE HOUSE SPARROW.

I am not entirely free from suspicions that by naturalizing the House Sparrow in this country, we have introduced a pest. It has always been regarded in Europe as a mischievous bird, but is tolerated because, like all the

Sparrow family, through granivorous for the most part, it destroys great quantities of grubs and insects during its breeding-season, which continues several months. Other circumstances that render the bird valuable are its domesticated habits, its permanent residence, and its proneness to live and multiply in the city as well as the country. The little Hair Bird, which is far more interesting and musical, is not a permanent resident, and cannot, from its habit of breeding in trees, become inured to the city. Perhaps, therefore, it need not be feared that the multiplication of the House Sparrows will diminish the number of our native birds. But I cannot, while dwelling on this subject, avoid the reflection that since our people are resolutely bent on the destruction of our native birds, it may be fortunate that there exists a foreign species of such a character that, like the white-weed and the witch-grass, after being once introduced, they cannot by any possible human efforts be extirpated. When all our native species are gone, we may be happy to hear the unmusical chatter of the House Sparrows, and gladly watch them and protect them, as we should, if all the human race had perished but our single self, welcome the society of orang-otangs.

I am pleased to learn that Dr. Brewer does not fear that their introduction will cause any evil to our native birds. If I were entirely satisfied of the correctness of his opinion, I should say welcome to the little intruders. They are at least valuable by affording amusement to children who are confined to cities, and who may watch and feed them where, if they were absent, but few other birds would be seen. But I will leave the House Sparrow to treat of a far more interesting family of birds, the common Domestic Pigeon.

## THE DOVE.

It is a matter of curiosity among naturalists that Doves and Pigeons, which are active and powerful on the wing beyond any known species, should have submitted so readily to domestication. Their power of wing and consequent capacity of providing food for themselves at great distances from their habitations must render them quite independent of any necessity of resorting to man's protection, like the gallinaceous birds. Yet they have probably been domesticated, like the common fowl, from immemorial time. The Dove is a bird which has been sacred in all ages as an emblem of constancy, while hardly a gallinaceous bird could be named that does not in its moral habits represent the political theory of free love. Ornithologists have lately removed the Dove into a separate family, reclassing it as distinct from all other birds. Doves are, in a wild state, very powerful on the wing; but, having small feet and legs, they are awkward and feeble walkers. The Goose is said to fly to a greater height than any other bird; but none can equal wild Pigeons in swiftness. This power of flight is of great service to them when foraging; for they can have a roost in Virginia and sally forth in any direction fifty miles to obtain a breakfast, and return sooner than the steam-cars could perform the journey in one direction.

The Dove, — the most amiable of birds, consecrated to some of the kindest virtues of the human soul, dedicated in ancient times to Venus, whose chariot was drawn by two Doves, — like a sweet maiden who neither flaunts nor glitters, but gains admiration solely by her innocence and her beauty, is very properly considered the symbol of purity and holiness. Holy Spirit and Heavenly Dove are, in the poetry of Christianity, synonymous expressions. The Dove, in Biblical Fable, that was sent out by Noah to determine the condition of the earth after this great captain and his family had become weary of navigating the Ark, brought back the olive-branch, which, like its feathered bearer, has ever since been regarded as the emblem of peace.

The Dove is more completely domesticated than the Quail could be under any circumstances. But it is almost exclusively granivorous, and is not so useful a bird as the Quail, flocks of which, if protected by providing them food and shelter, would frequent our orchards, and rid the trees entirely of the canker-worms by picking up the insects that generate them before they have climbed the tree. Mr. George W. Rice of West Newton has for several years past kept his apple-trees free from canker-worms by means of early chickens. He binds a raw cotton band round the tree very near the ground. Before the insects have time to creep over this obstacle, they are caught by the hens and chickens, so that not more than one in a hundred escapes.

Doves of all species seem to be very similar in their manners. Almost the only notes they utter are a gentle cooing, and if you scare one it does not scream, like other birds, but makes only a low moaning. Hence arose the reputation of the Dove for gentleness. Yet it is not without spirit or courage. When a boy I had a flock of thirty pigeons, all white. I watched them so attentively that I learned all their peculiar habits, the constancy of the mated female, the gallantry of all males toward unmated females, and the courage with which both sexes would defend their place and nest. I could distinguish each one of the flock from all the rest, and had a name for each. They were all black-eyed but one, and this one had a slight tinge of lilac upon its white feathers, and its eyes were light gray. The common slate-colored Pigeon has red eyes.

## THE TURTLE-DOVE.

The first wild bird I captured and tamed in my boyhood was taken from the nest of a Carolina Turtle-Dove. The nest was placed upon the horizontal branches of a small white pine about fifteen feet from the ground. It was made of slender twigs put together as carelessly as if they had fallen from some branches above, and were levelled, but not hollowed, by the parent birds. The nest contained a single white egg, more roundish in its shape than that of the common tame Pigeon. I took the young bird from the nest when it was nearly ready to fly. I fed it exclusively upon farinaceous food, and was successful in rearing it. It grew very tame, and behaved like the young of a domesticated Dove. It often flew away in quest of food and regularly returned, and was so docile as to sit upon my hand. I exchanged the bird, to gratify one of my schoolmates, for a volume of Peter Pindar's works, which I read over and over again with great delight, and a volume of President Monroe's Tour, which I used for kindling-paper.

After I had taken the bird from the nest I heard for more than a week the almost uninterrupted cooing or moaning of the parents, or one of them, upon an old white oak that stood in a field near my boarding-house, which was almost surrounded by woods. This oak was about a quarter of a mile from the nest, and it seemed as if the old birds had in some way or other a suspicion of the fact that the young one had been removed in this direction from the nest. To listen to the "mourning Dove" was a romantic incident that gave me so much satisfaction that it entirely absorbed all the sympathy I was disposed to feel for the bereaved parents. The young Dove was shot soon after I parted with it by one of the pioneers of Christian civilization, a Divinity Student.

Turtle-Doves are now rarely seen in New England, but they are common in other States. In this centre of enlightenment there is plenty of cant about mercy to birds and other creatures; there are whole encyclopædias of rhymes written about the "beautiful and innocent birds." But the rhymes and the cant go hand in hand with the snare, the gun, and strychnine; as the Bible and missionaries sail lovingly together with rum and gunpowder, to Africa and other regions of moral darkness, sent onward by the same persons and the same funds. There may be some desire in many hearts for the preservation of our birds; but it is with our sentimentalists as with our politicians, sentiment must give way to peas and strawberries as principle must give way to party and personal ambition. It is a remarkable fact that the possession of a single cherry-tree or one bed of strawberries will turn the most lachrymatory sentimentalist into a rabid exterminator of the feathered race.

### THE COCK.

I should be guilty of a great omission, if in my descriptions of interesting birds I were to say nothing of the common Cock, the true Bird of Morn in every country; the monitor who never fails to give the inmates of the house notice of the dawn of day. So intimately is this bird allied with the morning, that the dawn is always designated as the hour of cock-crowing. If he should cease hereafter faithfully to announce the earliest approach of day, we should look upon him as one who had lost the most remarkable trait in his character. But, like other birds that sing by night, he is often deceived by the light of the moon, when it rises past midnight, mistaking its beams for the promise of dawn.

The Cock is a bird of the East, and is by nature addicted to Eastern customs and habits. He is furnished with spurs with which he is expected to fight for the possession of as many females as he can procure by slaying his rivals. He knows no such feeling as an exclusive attachment to a single mate. He is a bird neither of sentiment nor principle. His crowing is but sound of triumph and exultation which is designed to notify all his brood of wives of his presence and of his power to defend them, and his defiance to other males who should venture to claim any one of the numerous members of his harem. His example has always been copied by the kings and sultans of the East. There is only this difference, — that the Cock obtains by his prowess what the sultan obtains by wealth and political authority, aided and countenanced by the deity whom he worships. But if Solomon was like Chanticleer in his customs, we might apply to him a quotation from the New Testament: "That even Solomon in all his glory was not arrayed like one of these."

The variety of plumage which is displayed by this bird in his domesticated state surpasses that of any known species. It is remarkable, however, that he has very few pure colors. He has no pure yellow, nor blue, nor crimson, nor scarlet, nor vermilion. But there is a brilliancy about these neutralized colors and there are fine contrasts in their arrangement giving splendor to certain varieties of this bird that cannot be surpassed. There are some which are pure white and others pure black. In these varieties the male and female differ but slightly in color. In other varieties, if the female is brown, the male is red; if the female is black, with neck-feathers grayish striped, the male is black, with neck and saddle feathers of a bright buff color. If the female is all gray, the male is gray, with neck and saddle white and tail black. Several of these contrasts are very beautiful. The long silken feathers of the neck and saddle distinguish the Cock from almost every other bird save the Pheasant. The Peacock, the Turkey, and the Guinea-Fowl are destitute of these marks.

### THE TURKEY.

The Turkey is not so interesting a bird as the Cock. He is neither so lively nor so courageous. His gobbling is not so musical as the crowing of the Cock, nor is it in any respect a sentinel sound. He resembles the Peacock in many ways, but does not equal him in beauty. But the wild Turkey is said to be in all respects more beautiful than the tame one. There was formerly some controversy respecting the American origin of this bird. Beside the whimsical Daines Barrington, many eminent naturalists supposed Africa to be its native country. Buffon, however, eloquently supported its claims to be considered an American bird. C. L. Buonaparte says, the first Turkey that garnished a feast in France was served up at the wedding banquet of Charles the Ninth in the year 1570. This was also the date of the general introduction of the Turkey into Europe as a domestic bird.

Dr. Franklin wrote a characteristic piece of humor on the substitution of the Turkey for the Bald Eagle as the emblematic representative of our country. The Bald Eagle he considers a bird of bad moral character, who gets his living by dishonest means. Like a robber he watches the Fish Hawk, and when he has caught a fish, pounces upon him and takes it away from him. Withal, he is a rank coward, and permits himself to be driven out of the district by the little Kingbird. He confesses, therefore, that he is not displeased that the figure is not recognized as a Bald Eagle, but looks more like a Turkey. The Turkey is a more respectable bird, and a true native of America. He is also, though a little vain and silly (and, as the Doctor expresses it, "not the worse emblem on that account"), a bird of courage, and would not hesitate to attack a grenadier of the British Guards who should invade his grounds with a red coat on.

Wild Turkeys were formerly not uncommon in the woods of New England. If any still remain they will not long escape the besom of civilization and progress. The Turkey will vanish with the Turtle-Dove and the Quail, and go where arithmetic and trigonometry have not yet mapped out the wilderness into auction-lots.

## THE GOOSE.

The Goose is truly a pastoral bird. Though it uses animal food, it lives more upon grain and by grazing, like cattle and sheep. It is not a sea fowl. It devours some insects, but does not take fishes, and resorts to the water chiefly at night, where it retires to rest, for security. It is the pastoral habit of the Goose that renders it so fit a subject for domestication. On the same account it is a better walker than the Duck, that passes the greater part of its time in the water, feeding upon the aquatic vegetables that grow in the shallows and upon such insects as are found among them. The Goose, notwithstanding the general habit among us of using its name as the superlative of folly, is an intelligent bird. The proverb "silly as a Goose" would be more correctly applied to a Hen or a Turkey.

The Goose has no special beauty of plumage. Its colors seldom vary from white and black and gray. The wild Goose of America greatly surpasses the common domesticated species in beauty, having some fine shades of green and purple on the black feathers of its long swan-like neck. Charles Waterton says of this species, which has been very generally domesticated in Great Britain: "There can be nothing more enlivening to rural solitude than the trumpet-sounding notes of the Canada Goose. They may be heard at most hours of the day and during the night. But spring is the time when these birds are most vociferous. Then it is that they are on the wing, moving in aerial circles round the mansion; now rising aloft, now dropping into the water, with such notes of apparent joy and revelling as cannot fail to attract the attention of those who feel an interest in the wildest scenery."

Wild Geese and other birds of the same family assemble, not in myriads, like Pigeons and Blackbirds, but in such limited flocks as admit of organization and geometric arrangement. Geese sometimes fly in a straight line; but more frequently make a triangular figure, that permits each one in the rear to see its leader. Some naturalists say that Geese fly to a greater height than any other bird; others say they are surpassed by Herons. They are often, however, at so great height that they may be heard, when nothing more of them than a black line can be seen. Before they alight upon the ground they form a straight line, probably without any purpose but from the habit of arranging themselves in a single rank and file when flying. Having taken their rest for a few hours, the sentinel gives the signal note, when they all rise again, form the same triangular group, and pursue their mysterious journey to a southern clime.

Naturalists are not agreed respecting the character of the leader of these flocks. Some believe that an old gander who has previously made the journey takes the lead. Others assert that each one of the flock takes his turn in being leader. It seems to me highly probable that neither of these assumptions is correct; but on the other hand, that the leadership is a matter of chance, except that the most powerful individuals would usually happen to place themselves at the head of the flock, being naturally the most active and vigorous, the first to rise from the ground and the swiftest to gain the foremost position. It is absurd to suppose that these birds in their migrations are directed by the knowledge and experience of a few older ones. Urged by a powerful impulse, if the old birds were all destroyed, the young flock, when the proper time arrived for their migratory flight, would proceed on their journey as instinctively as they would konk instead of crowing like a Cock.

### THE DUCK.

Ducks are by far the most beautiful of all aquatic birds in the colors of their plumage. Other genera of this family seldom show any hues except a various mixture of white and gray. The plumage of several species of the Duck is of many colors and finely variegated. This beautiful lustre is remarkable in the drake of the Mallard, of the Teal, and above all of the Summer Duck. Of the latter, both male and female are beautiful, and the species was named by Linnæus, on account of its beauty, *sponsa*, a bride. Its pendent crest of green and purple hanging from the back of its head; its neck of purple-crimson, changing in front to a glossy brown, speckled with white; its wings and tail of metallic green, changeable into blue and crimson,— its endless varieties, indeed, of changeable hues cause it to surpass in beauty all the birds of our woods and waters.

It is not often that we have an opportunity of watching for any considerable time the manœuvres of wild Geese or wild Ducks upon the water. We must observe the motions of the domesticated birds to learn those of the wild ones, making allowance for less dexterity, as the consequence of domestication. The flight and habits of the Duck are not less interesting or picturesque than those of the Goose. Their whistling flocks that pass frequently over our heads at different seasons always command our attention. Ducks live the greater part of the time upon the water, feeding upon the plants that grow around their edges and borders. Hence they prefer small ponds and inlets of the sea to the bay or harbor. But, like almost all other species of birds, the Duck and the Teal are rarely seen except in the remote lakes of the forest. These wild birds are allowed no peace and no security. I cannot see what is to prevent their utter extirpation from the American continent.

The Black Duck seems more nearly allied to the Mallard than to any American species. It has been repeatedly domesticated, and mixes with the Mallard, and the mixed offspring have none of the marks and qualities of hybrids. The drake of this species has not the beauty of the Mallard drake. Flocks of them are common in the autumn in some of our solitary inlets or near our harbors; and they formerly reared their young in Massachusetts. They have been driven away by gunners, and they now breed only in the northern parts of New England, especially near the lakes of Maine. Samuels found the nest of one on a low stump, that overhung a small spring on the side of a hill, a mile from any water. He says these nests are abundant all round Lake Umbagog. When the fresh ponds are frozen, the Ducks resort to the salt water, and are often seen, in flocks of considerable size, in our harbors and salt-water creeks in winter.

### THE SWAN.

If the Duck is the most beautifully arrayed of all aquatic birds, the Swan is certainly the most graceful and attractive when sailing upon the water. The Swan resembles the Duck more than the Goose in its feeding habits. It does not graze like the Goose, but takes its food from beneath the water, often probing to the bottom

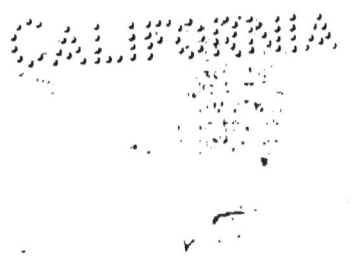

of shallow waters by means of its long neck, which seems designed for this purpose. Wild Swans associate in small flocks, separating in pairs during the breeding-season, and rising in large companies when the approach of winter warns them to seek a more genial clime. When they finally take their migrating flight, they divide themselves again into small flocks, and shape their course after the manner of Wild Geese.

# THE FLIGHT OF THE WOOD-NYMPHS.

ON the southern slope of a hill, nearly in the entrance of a valley, stood a rustic cottage inhabited by a plain industrious farmer and his family. The farm which was connected with the cottage was a beautiful intermixture of wood, tillage, and pasture; and, imbosomed in these natural groves, the glistening waters of a miniature lake gave animation to the landscape, and became a scene of rustic sport for many a youthful angler. In front of the cottage was an irregular grassy slope, extending down to the roadside, and wholly unenclosed. Through this natural lawn a narrow footpath, leading obliquely from the road to the doorstep, had been worn by the feet of passengers; tufts of wild shrubbery grew here and there about the rocks that projected from the surface of the soil, and the sweet fern diffused its odors within a rood of the cottage windows. In the evening a small herd of cows might be seen, quietly ruminating under a rugged old oak, that stood about thirty paces from the house.

In the month of May this place was a favorite resort for all the children of the village, on account of the multitude and variety of wild-flowers that grew there, and the many pleasant arbors afforded by the woods that overshadowed the borders of the lake. On these green hillsides they might often be seen weaving chains of the stems of dandelions, or stringing white and blue violets upon a thread, with which they made garlands and necklaces to add to their own simple apparel. Later in the season, old and young resorted hither, to gather berries that grew abundantly in these grounds and the neighboring pastures. Many a May-queen has been crowned with the trailing evergreens that abounded in all the wild lands, and covered the meadows with verdure in the depth of winter; and the children have returned home with baskets full of checkerberries and garlanded with early spring flowers.

There was something about the whole aspect of this place that was unaccountably delightful. Every one who visited it felt inspired with a mysterious sense of cheerfulness and pensive delight, that could hardly be explained, as there were in the same region many magnificent country-seats, with highly ornamented grounds, that failed in awakening any such emotions. Here nothing had ever been done to add a single ornament to the face of nature, but in all parts of the landscape there was a beauty that seemed unattainable by art. It became evident at last that these groves and pastures must be the residence of the rural deities, who, by their invisible presence, inspired every heart with those delightful sentiments which, though not entirely unfelt on earth, are well known only in Paradise. It was the presence of these deities that yielded the place its mysterious charms. It was the naiad who gave romantic melody to the fountain that bubbled up from the mossy glen in the hillside, and spread the hue of beauty over the solitary lake in the valley; and the dryads, or wood-nymphs, that caused these woodland arbors to rival the green retreats of Elysium.

In these rural solitudes were assembled all those little harmless animals, which by their motions and frolics serve to give life to the inanimate scenes of nature. Here were not only all the familiar birds that delight in the company of man; but all the less familiar species that love to chant their wild melodies in the hiding-places of the solitary echoes, might also be heard in the season of song.

The red-winged starling, long exiled from our villages, still uttered his melancholy ditty among the willows in the valley, and wove his nest among the tall rushes that rose out of the water. The ruff-necked grouse beat his muffled drum in the adjoining forest, and the hermit-thrush poured forth his indescribable strains, like some voice that had wandered from the groves of Idalia. Even in the depth of winter the hearts of the farmer and his family were cheered by a multitude of merry voices, that seemed to be peculiar to the place.

This charming spot soon became celebrated in all the country around for its romantic beauties; and it was eagerly coveted by many people of wealth who were seeking a place of rural retirement. The cottager who had lived here ever since his birth regarded it with affection and reverence, as his own paternal homestead. But there are not many who can resist the temptation of gold to make a sacrifice either of principle or affection, and the rustic possessor of this little farm was not one of them. He sold it to a man of wealth and cultivated taste, whose wife and daughters were unaffected lovers of nature, and who were delighted with the idea of occupying a place that was celebrated as the resort of the wood-nymphs and other deities of the groves. The new proprietor determined to adorn and improve it to the utmost extent. He resolved that the decorations of the modern landscape art should be added to the advantages it had derived from nature; the beauties of other climes should be engrafted upon it, and the whole work should be crowned with the best efforts of the sculptor and the architect.

In accordance with these plans, the work of beautifying and improving the place was commenced. Standard English works on landscape gardening were consulted; the great Italian painters were studied for hints which Nature is supposed to communicate only through their medium, and Brown and Repton guided the taste of the improver in all his operations. The rustic cottage was removed to a distant spot, and a splendid Italian villa was erected in the place of it. No labor nor money was spared in the effort to give it all the external and internal finish which would be needful to adorn a palace. Every piece of work was tasteful and correct; no counterfeit imitations of valuable ornaments were allowed; and when the edifice was completed, the most scientific architect could find no fault with it. It stood forth proudly on the brow of the hill, one of the masterpieces of villa architecture.

The elegance of the mansion made it the more apparent that the grounds must be improved, that the appearance of nature might harmonize with the work of the architect. On the grassy slope that fronted the cottage, there were occasional projections of the rock that was buried underneath the soil, and around them various species of wild shrubbery had come up in many a tufted knoll. These prominences were split off, and covered with loam, and the whole surface was graded into a beautifully even and rounded lawn. The wood-anemone, the mouse-ear, and the saxifrage no longer spangled the grassy slope in early spring, nor the aster nor the golden-rod stood there to welcome the arrival of autumn. But tulips grew proudly in a fanciful border of spaded earth under the side windows in the opening of the year, and verbenas, portulaccas, and calceolarias outshone all the native summer beauties of the landscape.

Surrounding the fields that adjoined the cottage was an old stone-wall, gray with lichens and covered with numerous wild vines that had clustered round it, as the ivy intwines itself round the walls of ruined castles and abbeys in the Old World. The clematis overshadowed it with flowers and foliage in summer, and with its beautiful silken down in the fall of the year; and the celastrus grew with it side by side, offering its honeyed flowers to the bee, and its scarlet bitter-sweet berries to the hand of the simpler, or to the famishing winter birds. Among this vinery the summer warblers built their nests; and numbers of them were revealed to sight when the foliage was swept away by the late autumnal winds.

The ladies of the mansion would not readily consent to the removal of this old stone wall, with its various rustic appurtenances, which seemed to them a part of the original charms of the place; but they were soon convinced that the villa ought not to stand in the midst of such shabby "surroundings." They were plied with arguments drawn from the works of men who had studied nature in the galleries of art and through the medium of canvas, and were persuaded to believe that the principles of English landscape-gardening must never be sacrificed to the crude notions of a poetic mind. The ladies gave up their impulses in favor of the cold rules of professional taste. The stone wall was removed; the wild rose and the eglantine were destroyed; the flowering shrubs that formed, on each side of it, a glistening row of bloom and verdure, were rooted up; a neat paling fence was erected as a temporary boundary, and a hedge of buckthorn was planted all around the old pasture!

The lawn in front of the mansion was enclosed by an ornamental fence, and the narrow footpath that led up to the rude doorstep of the cottage, meeting in its course an occasional tuft of spiræa and low laurel, gave place to a neatly gravelled walk, four feet six inches wide, and shaped into a graceful serpentine curve. The enclosure was filled with exotic shrubbery; and silver maples, silver poplars, and silver firs stood at proper distances, like sentinels to guard the portals of this temple. The grounds were likewise embellished with statuary, and large marble vases, holding some flaunting exotic, stood in their assigned positions.

Two years had not elapsed before the design of the improver was completed, and the whole aspect of the place was changed, as if by enchantment. The rustic cart-paths that led over the hills and through the woods and valleys were widened and covered with a neat spread of gravel, and all their crooked outlines were trimmed into a graceful shape. An air of neatness was apparent in every direction. The undergrowth of the wood was removed, certain misshapen trees were cut down, and all rubbish was taken away that could afford a harbor for noxious insects or mischievous quadrupeds. The lake that was embroidered with alders, swamp roses, button-bushes, the fragrant clethra, and the drooping andromeda, was improved by the removal of all these useless plants, and gravel and loam were carted down to its edges, which were then covered with soil and sowed with grass seed, to afford a neat and lawn-like appearance to the grounds, and to visitors a firm foundation for their feet. The frequent tufts of shrubbery that gave a ragged look to the pasture were likewise removed, and the whole was planted with the most approved grasses.

Not many rods from the cottage was a natural fountain that bubbled up from a subterranean source in the hillside, from which the farmer irrigated the greater part of his lands. It was a true rustic fountain, girded on one side by steep fern-clad rocks, and overshadowed by the gnarled and twisted branches of the tupelo, one of the most grotesque and beautiful trees in the forest. From this fountain issued a rivulet, which was conducted along the declivity, until it poured its waters into a wooden trough, and formed a watering-place for the cattle. These objects were altogether too rude to be admitted as a part of the map of improvements. The bed of the fountain was excavated into a deep and spacious reservoir, and from this a pipe was carried along under ground to the front yard, where it terminated in a *jet d'eau*, that issued from a marble basin, and threw up a wide and graceful spray.

The inmates of the villa were charmed with the result of these operations. There was an air of elegance and "high keeping" about the grounds, that corresponded judiciously with the splendor of the villa and its outbuildings. No wild bushes were left in straggling tufts, to suggest the idea of poverty or negligence on the part of the proprietor; and the pasture, which was full of a great variety of wild plants or weeds, was repeatedly ploughed and pulverized to destroy them, and afterwards "laid down" to legitimate English grasses. The dandelion and buttercup were no more to be seen in the spring, or the rank hawkweed in the autumn; through this lawn neat gravel-walks were made, that visitors might stroll there in the morning without being wet by the dews. Many of the slopes were provided with marble steps, and here and there, in the centre of a clump of firs, were erected marble statues to emblemize the rural deities.

But where stands the idol, there we may not feel the presence of the deity. In vain do we strive to compensate Nature, when we have despoiled her of her original charms, by calling in the aid of the sculptor, whose lifeless productions serve only to chill the imagination that might otherwise revel among the wizard creations of poetry. The images of Ceres, of Galatea, or of the heavenly huntress were not attractive to the beings whom they were intended to represent. The naiad no longer sat by her fountain which was held in a marble basin, and sent up its luminous spray, in the midst of the costly works of art. The dryads had forsaken the old wood, whose moss-grown trees were deprived of their variegated undergrowth, and of the native drapery that hung from their boughs. They wept over the exiled birds and the perished flowers of the wildwood, and fled sorrowfully to some new and distant haunts. The nymphs who used to frequent these shady retreats had also fled. Woods, groves, hills, and valleys were all deserted; and the cold, lifeless forms that were carved out of marble stood there alone, the mere symbols of charms that no longer existed.

The village children who formerly assembled here to gather bouquets of wild roses, red summer lilies, and the sweet-scented pyrola, that grew up like a nun under the shade of the deep woods, came often since the improvements, but searched in vain for their favorite flowers. They no longer saw the squirrel upon the tree or the nest of the sparrow upon the vine-clad wall. The grounds, that seemed once to belong to them as well as to their rustic proprietor, now displayed something in their aspect that made them feel like intruders, as soon as they set foot within their borders. These old woods and pastures, now that they were metamorphosed into park and lawn, had lost their charms for them, and they turned away with sadness, when they thought of the delightful arbors that would shelter them no more.

But the children were not the only sorrowers. The ladies of the mansion were grieved when they found that the rural deities had fled from the very objects which were erected for their shrines. The cause of their flight was a problem they could not explain. Why would they no longer dwell in their ancient abodes that seemed now so much worthier the residence of beings of a superior nature? Could not the beautiful green lawn that had taken the place of the weedy pasture, nor the commodious park which was once a tangled wood; nor the charming flowers of all climes which had been substituted for the inferior wild-flowers, nor the marble fountain with its graceful spray, nor the neat-spread gravel-walks, induce them to remain? More than all, could not the beautiful statuary that represented them in material shape please them and retain them in their ancient haunts?

At length they began to suspect that there was a too entire absence of rustic scenes and objects in their present arrangements; and forthwith, to appease the deities, rustic arches and bowers, made of rude materials, were erected and placed in different parts of the grounds. A summer-house was built of the rudest of logs, shingled with the rough bark of trees, and rocks were introduced for seats and covered with mosses. Fences were constructed in similar style, and various other rude devices were executed and distributed in a fanciful manner over the face of the landscape. But not even the shaggy goat-footed Pan would acknowledge any such thing for an altar. No such objects could be made to accord with the "high keeping" of the grounds, nor could they give an air of rusticity to the scenes that were so elaborately ornamented. They were mere pieces of affectation; blotches upon the fair surface of beauty, that served no other purpose but to add deformity to the unique productions of art.

One day, as the ladies were strolling pensively along their accustomed paths, lamenting that nothing could be done to appease the divinities whom they had offended, they discovered in a little nook, under a cliff that projected over a rude entrance into the wood, a slab of weather-stained slate, resembling a headstone. Observing that it was lettered, they knelt down upon the green turf and read the following

### INSCRIPTION.

In peaceful solitudes and sylvan shades
That lure to meditation; where the birds
Sing all day unmolested in their haunts,

And the rude soil still bears the tender wilding, —
There dwell the rural deities.  They love
The moss-grown trees and rocks, the flowery knoll,
The tangled wildwood, and the bower of ferns.
They fill each scene with beauty, and they prompt
The echoes to repeat the low of herds
And bleat of tender flocks.  The voice of him
Who drives his team afield ; the joyous laugh
Of children, when on pleasant days they come
To take from gentle Spring her gift of flowers,
Are music to their ears.  All these they love ;
But shun the place where wealth and art have joined
To shut out Nature from her own domains,
Or dress her in the flaunting robes of fashion.
Wouldst thou retain them ? — keep a humble heart,
Nor in their temples seek to show thy pride,
Or near their altars to parade thy wealth ;
Then may they come and dwell with thee as once
With simple shepherdess and rural swain.

# CELESTIAL SCENERY.

THE system of Nature is attended with so many circumstances that mar our happiness, that Nature has benevolently spread every scene with beauty that shall serve by its exhilarating influence to lift us above the physical evils that surround us and render us half unmindful of their presence. Hence beauty is made to spring up, not only in the field, in the wilderness, and by the wayside, by the sea-shore and among the hills, but it is spread in gorgeous spectacles upon the heavens in the infinitely varied forms and arrangements of the clouds, and in their equally beautiful lights, shades, and colors. The man of feeling and culture, therefore, who takes pleasure in surveying the beauties of a terrestrial landscape feels no less delight in contemplating the scenery of the heavens. Every morning, noon, and evening affords him scenes always charming and never tiresome, being as changeable and evanescent as they are brilliant and beautiful.

I have ever been at a loss to explain why we are more agreeably affected with the appearance of sunshine on a circumscribed part of the landscape, while we ourselves are enveloped in shadow, than when the whole space is illuminated. In this case the circumstances that cause us to look with pleasure upon the raging of a tempest, while we are in a comfortable shelter, seem to be reversed. The two facts, however, do not involve any inconsistency. In the first case, we are amused in a comfortable lookout, with the movements of a tempest, — the whirling of the clouds, the falling of the rain, the flashes of lightning, and the roar of thunder. In the other case, the sunshine makes an agreeable picture. It affords a view that cannot so well be seen in shadow or where the sunlight is equalized over the whole prospect. It is set apart from the remainder like an island in a lake. But, above all, the tract of country thus illuminated, while our standing-place is shaded, becomes to the imagination a celestial view. The scene in sunshine is made a part of the heavens; and the mind is exhilarated on beholding a scene in our earthly landscape exalted as it were to the skies.

The moon has always been a favorite theme of the poets. Her course in the heavens has in all ages been marked with interest, and her form and phases watched with delight. We associate her beams with serenity and peace. Her very aspect breathes of purity and holiness. How gloomy and lonely would be the night without her presence, except with the knowledge that she will soon reappear to bless the earth with her beams. Nature seems to have regarded this luminary as indispensable to the moral wants of rational beings; but in this, as in other cases, she has been cautious of prodigality. I have often thought that two or three moons would be less delightful than the solitary orb that guards our night. As the moon is not needful, like the sun, for the existence of light and life, but is rather one of the luxuries of Nature, her light is more beautiful than useful, while we do not suffer from her occasional absence.

Lovers have always been charmed by moonlight, that accords so well with seclusion and tender sentiment. "A fair face looks yet fairer under the light of the moon, and a sweet voice is sweeter among the whispering sounds of a summer night." This remark of Walter Scott describes what almost all persons have felt. The beauty of a lovely countenance seems to partake of a more spiritual character in the mellowing light of the moon; for this luminary brings us nearer heaven than the sun, by our sequestration in the darkness that surrounds us.

The moon is regarded by those who are melancholy or affected with grief as a heavenly sympathizer. They welcome her soft and pensive light, to divert the soul from the misery of its own thoughts and to breathe into it that serenity which pervades her countenance. To a religious mind this fair orb seems very properly a heavenly gift intended for the refreshment of the soul, especially as the physical benefits conferred upon us by the moon are not apparent to reason and observation. Hence we hail this luminary when ascending from the misty verge of the horizon as a fair messenger of heaven, and we are inclined to pay to her the homage of the soul as to a living deity.

In ancient mythology the moon is a serene goddess enthroned among the constellations, — the daughter of Jove, the heavenly huntress, the chaste Luna, the incorruptible Diana. She is the embodiment of all purity, of all blessedness, of tranquillity, and of hallowed and constant affection. She is the guardian of innocence, the protectress of virtue, the light of heaven in darkness, the guide of hope in despondency, the soother of grief, and the source of that tranquil inspiration that comes from peaceful themes and pleasant recollections. Her course in heaven is the path of peace, and her light is the same that dwells in the souls of heroes and inspired bards.

The light of the moon guides our steps without clearly revealing our presence to others. She is therefore the symbol of benevolence, yielding to the fugitive the means of finding safety and granting him her light for his own deliverance while it is insufficient to betray him. It is for this that the moonlight is so dear to one who seeks seclusion in an hour of night that cannot be devoted to rest. Twilight partakes of the same quality, but in a less degree. It is more diffused, having no shadows under which an object might be concealed, while its light is of less avail to the wanderer. Moonlight, above all other kinds of light, is therefore the comfort of those who seek concealment that they may enjoy an hour of freedom; who would meditate awhile without interruption, or hold sacred intercourse with a companion whom it would be imprudent to meet under the broad light of day. Hence those who are cheerful and those who are depressed, and those who are anxious and afflicted, hail the light of the moon with gladness and her placid countenance with veneration.

To comprehend the full glory and beauty of moonlight, it must be seen at one time upon the calm surface of a lake and again upon the ruffled tide of the ocean. In one case, it images its own serenity upon the placid mirror; in the other, it forms a beautiful contrast with the agitated state of the waves,—the peace of heaven opposed to the distracted condition of the inhabitants of the earth.

The light of the moon is seldom iridescent. Her radiance is mostly of a pure whiteness. When her light falls upon the clouds they assume no gorgeous colors; they display that silvery light only that symbolizes purity. She often wears a corona, and gives thereby a prophetic signal to the laborers, who bless the token as the omen of refreshing showers. This corona is projected upon the highest clouds, and indicates the prevalence of moisture in the upper part of the atmosphere. No such appearance is impressed at any time upon the lower clouds. The lunar rainbow is a beautiful but rare phenomenon, which I have seen only once in my life.

Though I have treated of the moon as an object of more passionate contemplation than the sun, the effects of sunlight are infinitely more glorious and beautiful. The sun, being too intensely bright to be viewed by the naked eye, must be adored in its effects, — in the beautiful tintings of sunrise and sunset, in the silvery lustre of the clouds at noonday, in the halo that surrounds his disk and gives warning of a tempest, and in the rainbow that announces the end of the storm. It is in these celestial phenomena that we behold the beauty of colors in their highest degree. Of all that is beautiful on earth, there is nothing that equals the beauty of the sun's rays upon the clouds. There is nothing so exhilarating to the mind, or that conveys such a vivid consciousness of the existence of something purer and more divine than our life in this world.

# FEBRUARY.

WHEN we consider the general sameness of winter's aspects, we need not marvel that among the works of landscape-painters there are but few pictures of winter. These few have generally represented some domestic scene, — a cottage with its roof covered with snow; cattle standing in a warm shelter in the barnyard; poultry huddled in a sunny corner; and children hastening toward their homes. Among the designs of Thomas Bewick there is only one winter scene, and this has served as the original from which all later ones have been copied or imitated. It represents a traveller with a pack on his shoulders, trudging over a trackless region of snow-covered ground, accompanied by his dog. He makes his way, not like a man who is enjoying his walk, but as one beset with dangers and thinking only of gaining his journey's end. The sun shines coldly upon him, and the wind causes him to bend to its blast. The naked trees frown upon him, his lengthened shadow seems like the ghost of Winter forever haunting his sight, and his dog looks up to him piteously and seemingly anxious to know his master's thoughts.

Whenever we ramble in winter we can readily understand why the naturalist, who studies individual objects, should find but few attractions in a winter's walk; but it is not so clear why the painter, whose principal purpose is to observe aspects, should be uninterested. If we are inclined to indulge in meditation, no other season is so favorable to it. In the agreeable monotony of a snow-scene, there is but little to divert attention from our thoughts. We can find enough to employ our observation; but there is less than at other seasons that forces itself upon our attention. We can leave ourselves at any time, to examine a remarkable object or to view a charming scene.

He must have an eye that is insensible to grandeur and a mind that is incapable of appreciating the sublimity of landscape who would say that Nature is destitute of charms in the month of February. It is true that the variegated surface of brown and white that characterizes a winter prospect, though it be here and there diversified with a knoll of evergreen-trees that lift their heads as it were in triumph above the snows, will not compare with the interminable verdure of summer or the magnificence of forest scenery in autumn; yet there is a quiet sublimity that pervades all Nature — hill, field, and flood — at this season, which almost reconciles one to the temporary absence of summer flowers and spicy gales.

I am no lover of cold weather, and feel more contented when the sultry heats of summer oblige me to seek the refreshing breezes beneath a willow-tree on the banks of the sea-shore, than when the cold blasts of winter drive me within doors or force me to mope in a sunny nook in the forest. But there are days in winter, when the wind is still and mild, which are attended with pleasant sensations seldom experienced even in the month of June. Whether the delightful influence of this serene weather arises from a physical cause, or whether it is the result of contrast with the cold that has kept us half imprisoned for many weeks, I cannot determine. But when I review the rural rambles of former years, my winter walks on these delightful days will always crowd most sweetly and vividly upon my memory.

In winter the mind possesses more sensibility to rural

charms than during the seasons of vegetation and flowers. A long deprivation of any kind of pleasure increases our susceptibility and magnifies our capacity for enjoyment. Thus we may become indifferent to the warbling of birds in the summer, while we are forming a habit which, after the long silence of the wintry woods, shall cause the melodies of spring to yield us the greatest delight. After the confinement of winter we are keenly alive to agreeable impressions from all rural sights and sounds. Then does the sight of a green arbor in the woods or a green plat in a valley affect us as I can imagine a weary traveller to be affected on suddenly meeting an oasis in a desert. The melancholy that attends a ramble in the autumn has passed from us, and we now come forth, during the sleep of vegetation and in the general hush of animated things, with some of the gladness that inspires the mind when the little song-sparrow first sings his prelude to the general anthem of Nature. Some blessing comes from every sacrifice, and some recompense for every privation. Thus the darkness of night prepares us to welcome with gladness the dawn of a new morning. The charm of life springs from its vicissitudes, and we are capable of no new enjoyment until we have rested from pleasure.

When the earth is covered with snow that has grown hard enough to bear our footsteps without sinking into the drifts, I have often taken advantage of one of the serene days of winter to ramble in the woods. Every pleasant rural object I then behold affords me as much pleasure as I should derive in summer from all the charms of landscape united in one view. The snow lies in scattered parcels over the earth, that serves to variegate the scene and to render it more pleasing to the sight than the dull monotonous brown which the landscape wears at this season, when there is no snow.

Every sound I hear in the woods at such a time is

music, though it be but the cowbell's chime, the stroke of the woodman's axe, or the crash of some tall tree just falling to the ground. Sometimes, during this period of calm sunshine, the squirrels will come forth from their retreats and in the echoing silence of the woods we may hear their rustling leap among the dry oak-leaves, their occasional chirrup, and the dropping of nuts from the lofty branches of the hickory. There is music in all the echoes that break the stillness of the hour; in the cawing of crows, the scream of jays, or the quick hammering of the woodpecker upon the hollow trunk of some ancient standard of the forest..

The mild serenity of the weather, the fresh odors that arise from thawing vegetation, the beautiful haze that surrounds the horizon, reflecting all the colors of the rainbow, the lively chattering of poultry in the farm-yards, the bleating of flocks and the lowing of kine, an occasional concert of crows in the neighboring wood, the checkered landscape of snow-drifts rising out of the brown earth and gleaming in the sunshine, and the soft hazy light that glows from distant hills and spires, — all these rural sights and sounds affect us with a pleasure not surpassed by that which is felt at any time or season. Now and then, amidst all this harmonious medley, as if to remind us of the coming delights of spring, a solitary song-sparrow, prematurely arrived from the south, will tune his little throat and sing from some leafless shrub his first salutation of reviving Nature.

Among the attractions of winter scenery I must not omit the frostwork upon the windows, which has been so often used by poets to emblemize the hopes of youth. All vegetation in summer presents not a greater variety of forms than we may behold in these beautiful configurations. The mornings which are most remarkable for this curious pencil-work are such as follow á very cold night after mild and thawing weather on the preceding day. Nothing in the world seems so much like the effects of enchantment. The pictures made by the frost upon our window-panes are a part of the domestic scenery of winter; but their origin and progress form a curious study. It is remarkable that this deposit of frost resembles in structure and development the formation of clouds in clear weather in the upper region of the heavens. The clouds usually display more beauty of form in winter and in very dry weather, because the arid state of the atmosphere is favorable to their delicate organization. Hence the most beautiful clouds are those which are highest above the earth's surface, where the air contains but very little moisture. The same principle affects the formation of window-frost. The air of the room when only slightly charged with vapor projects the most delicate and beautiful figures on the windows.

The first deposit on the window-glass, when the weather is very cold and the air of the room moist, is a thin iridescent film resembling that produced by oil spread upon the surface of still water. This iridescence vanishes at the moment when the film begins to change into a crystallized surface. Immediately there appears in the place of it a collection of little *flocculi*, — a sort of constellation of minute snow-flakes, without any formal arrangement. These, as they increase from the moisture of the room, slowly assume a feathery organization, with more or less geometrical beauty, according as the deposit is made from air that is lightly or heavily charged with dampness. The less the moisture in the air of the room, if there be a sufficient quantity, and the colder the air outside if the inner air be not much above freezing-point, the finer and more beautiful are these configurations. Hence the windows of a sleeping-room, if not occupied by more than two or three persons, are more delicately frosted on a cold morning than those of a cooking-room where the moisture is precipitated so rapidly upon them as to mar their arrangement.

There is no season or month without its peculiar beauties. They are distilled like dew from heaven, and cover all places. They are scattered over the greensward in spring and summer, upon the forest in autumn, and in winter they are spread over the earth with the whiteness of snow and precipitated in frost upon the trees and upon our window-panes. At all times and seasons may we look upon these wonders and beauties that attract our sight in the least as well as the greatest operations of the Invisible Artist.

# INDEX.

## A.

| | | PAGE |
|---|---|---|
| Acadian Owl | *Strix Acadica* | 282 |
| American Goldfinch | *Fringilla tristis* | 14 |
| American Linnet | *Fringilla purpurea* | 16 |
| Angling | | 228 |
| Anthem of Morn | | 112 |
| April | | 72 |
| August | | 235 |

## B.

| | | |
|---|---|---|
| Baltimore Oriole | *Icterus Baltimore* | 86 |
| Bank-Swallow | *Hirundo riparia* | 251 |
| Barn-Swallow | *Hirundo rufa* | 248 |
| Bee-Martin | *Muscicapa tyrannus* | 255 |
| Birds of the Air | | 248 |
| Birds of the Farm and the Farmyard | | 419 |
| Birds of the Garden and Orchard, No. I. | | 5 |
| "     "     "     "     " II. | | 43 |
| "     "     "     "     " III. | | 83 |
| Birds of the Moor | | 333 |
| Birds of the Night | | 277 |
| Birds of the Pasture and Forest, No. I. | | 118 |
| "     "     "     "     " II. | | 164 |
| "     "     "     "     " III. | | 202 |
| Birds of the Sea and the Shore | | 391 |
| Birds of Winter | | 364 |
| Bittern | *Ardea minor* | 340 |
| Black Duck | *Anas obscura* | 430 |
| Bluebird | *Ampelis sialis* | 49 |
| Blue-Jay | *Corvus cristatus* | 377 |
| Bobolink | *Icterus agripennis* | 46 |
| Brigadier | *Vireo gilvus* | 43 |
| Brown Creeper | *Certhia familiaris* | 375 |

454                    INDEX.

## C.

|  |  | PAGE |
|---|---|---|
| Calculations | | 303 |
| Canada Goose | Anser Canadensis | 428 |
| Carpenter Bird | Picus principalis | 368 |
| Catbird | Turdus felivox | 169 |
| Cedar-Bird | Bombycilla Carolinensis | 90 |
| Celestial Scenery | | 442 |
| Chewink | Fringilla erythrophthalma | 121 |
| Chickadee | Parus palustris | 371 |
| Chimney-Swallow | Hirundo pelasgia | 254 |
| Clapper-Rail | Rallus crepitans | 337 |
| Cliff-Swallow | Hirundo fulva | 249 |
| Clouds | | 311 |
| Cock | Phasianus gallus | 424 |
| Cowbird | Icterus pecoris | 204 |
| Crane | Ardea Herodias | 342 |
| Crow | Corvus corone | 379 |
| Cuckoo | Cuculus Americanus | 202 |

## D.

| | | |
|---|---|---|
| December | | 384 |
| Dove | Columba | 421 |
| Downy Woodpecker | Picus pubescens | 373 |
| Drought | | 186 |
| Duck, Common | Anas boschas | 429 |

## E.

Early Flowers . . . . . . . . . . . . . . . . . 20

## F.

| | | |
|---|---|---|
| Facts that prove the Utility of Birds | | 415 |
| February | | 447 |
| Field and Garden | | 96 |
| Flicker | Picus auratus | 129 |
| Flight of the Wood-Nymphs | | 432 |
| Flowerless Plants | | 176 |
| Flowers as Emblems | | 131 |
| Foraging Habits of Birds | | 241 |
| Flowers of Autumn | | 267 |

## G.

| | | |
|---|---|---|
| Golden-crowned Thrush | Turdus aurocapillus | 124 |
| Golden Robin | Icterus Baltimore | 86 |
| Golden-winged Woodpecker | Picus auratus | 129 |

INDEX. 455

| | | PAGE |
|---|---|---|
| Goose | Anser | 427 |
| Green Warbler | Sylvia virens | 125 |
| Grosbeak, Rose-breasted | Fringilla Ludoviciana | 129 |
| Ground-Robin | Fringilla erythrophthalma | 121 |
| Gull | Larus argentatus | 397 |

H.

| | | |
|---|---|---|
| Habits of Birds, Changes in | | 329 |
| Haunts of Flowers | | 56 |
| Hair-Bird | Fringilla socialis | 12 |
| Hairy Woodpecker | Picus villosus | 374 |
| Hemp-Bird | Fringilla tristis | 14 |
| Hermit-Thrush | Turdus solitarius | 164 |
| Heron, Blue | Ardea Herodias | 342 |
| Hibernation of Swallows | | 262 |
| House-Sparrow | Fringilla domestica | 419 |
| House-Wren | Troglodytes fulvus | 52 |
| Humming-Bird | Trochilus colubris | 259 |

I.

| | | |
|---|---|---|
| Indigo-Bird | Fringilla cyanea | 92 |

J.

| | |
|---|---|
| January | 408 |
| July | 191 |
| June | 147 |

K.

| | | |
|---|---|---|
| Kingbird | Muscicapa tyrannus | 255 |
| Kingfisher | Alcedo alcyon | 394 |

L.

| | | |
|---|---|---|
| Lark, Meadow | Sturnus Ludovicianus | 88 |
| Log-Cock | Picus pileatus | 368 |
| Loon | Colymbus glacialis | 398 |

M.

| | | |
|---|---|---|
| March | | 31 |
| Marsh-Wren | Troglodytes brevirostris | 54 |
| Maryland Yellow-Throat | Sylvia trichas | 127 |
| May | | 105 |
| Meadow-Lark | Sturnus Ludovicianus | 88 |
| Mocking-Bird | Turdus polyglottus | 290 |
| Music of Birds | | 1 |

## N.

| | | |
|---|---|---|
| Night-Hawk } Night-Jar } | *Caprimulgus Virginianus* | 287 |
| November | | 354 |
| Nuthatch | *Sitta Carolinensis* | 376 |

## O.

| | | |
|---|---|---|
| October | | 324 |
| Old Houses | | 399 |
| O'Lincon Family | | 48 |
| Oven-Bird | *Turdus aurocapillus* | 124 |
| Old Roads | | 350 |
| Owls | | 278 |

## P.

| | | |
|---|---|---|
| Peabody-Bird | *Fringilla albicollis* | 18 |
| Pewee | *Muscicapa nunciola* | 256 |
| Picturesque Animals | | 138 |
| Plea for the Birds | | 157 |
| Plumage of Birds | | 78 |
| Preacher | *Vireo olivaceus* | 45 |
| Protection of Birds | | 197 |
| Purple Finch | *Fringilla purpurea* | 16 |
| Purple Grackle | *Quiscalus versicolor* | 208 |
| Purple Martin | *Hirundo purpurea* | 252 |

## Q.

| | | |
|---|---|---|
| Qua-Bird | *Ardea discors* | 341 |
| Quail | *Perdix Virginiana* | 210 |

## R.

| | | |
|---|---|---|
| Red-breasted Woodpecker | *Picus Carolinus* | 368 |
| Red-headed Woodpecker | *Picus erythrocephalus* | 374 |
| Redstart | *Muscicapa ruticilla* | 123 |
| Red-Thrush | *Turdus rufus* | 172 |
| Red-winged Blackbird | *Icterus phœniceus* | 206 |
| Robin | *Turdus migratorius* | 83 |
| Rocks | | 25 |
| Rose-breasted Grosbeak | *Fringilla Ludoviciana* | 129 |
| Ruffed Grouse | *Tetrao umbellus* | 212 |
| Ruins | | 296 |

## S.

| | | |
|---|---|---|
| Sand-Martin | *Hirundo riparia* | 251 |
| Saw-Whetter | *Strix Acadica* | 282 |

INDEX. 457

| | | PAGE |
|---|---|---|
| Scarlet Tanager | Tanagra rubra | 128 |
| Screech-Owl | Strix Asio | 283 |
| September | | 272 |
| Simples and Simplers | | 214 |
| Singing-Birds | | 37 |
| Snipe | Scolopax Wilsonii | 335 |
| Song-Sparrow | | 6 |
| Sounds from Animate Nature | | 318 |
| Sounds from Inanimate Nature | | 344 |
| Speckled Creeper | Certhia maculata | 123 |
| Spotted Tattler | Totanus macularius | 395 |
| Summer Yellow-Bird | Sylvia citrinella | 93 |
| Swan | | 430 |
| Swamp-Sparrow | | 119 |

T.

| Tattler | Totanus macularius | 407 |
|---|---|---|
| Testimony for the Birds | | 360 |
| Titmouse, Black-capped | Parus palustris | 371 |
| Turkey | Meleagris gallipavo | 426 |
| Turtle-Dove | Columba Carolinensis | 423 |

U.

| Upland Plover | Totanus Bartramius | 397 |

V.

| Veery | Turdus Wilsonii | 167 |
|---|---|---|
| Vesper-Sparrow | Fringilla graminea | 10 |
| Vireo | | 43 |
| Virginia Rail | Rallus Virginianus | 336 |

W.

| Water Scenery | | 66 |
|---|---|---|
| Whippoorwill | Caprimulgus vociferus | 285 |
| Why Birds sing in the Night | | 308 |
| Wilson's Thrush | Turdus Wilsonii | 167 |
| Winter-Wren | Troglodytes hyemalis | 54 |
| Woodcock | Scolopax rusticola | 333 |
| Wood-Sparrow | Fringilla pusilla | 120 |
| Wood-Swallow | Hirundo bicolor | 250 |
| Wood-Pewee | Muscicapa virens | 258 |
| Wood-Thrush | Turdus melodus | 166 |

Cambridge : Electrotyped and Printed by Welch, Bigelow, and Company.

www.ingramcontent.com/pod-product-compliance
Lightning Source LLC
Chambersburg PA
CBHW051236300426
44114CB00011B/760